Theories and Techniques
in
Vegetation Analysis

ROLAND E. RANDALL

Theory and Practice in Geography

OXFORD UNIVERSITY PRESS · 1978

Contents

1 Introduction

Vegetation is an extremely important element of the landscape. As such it deserves attention from the geographer who, in the past, has tended to devote his attentions either to the earth science (landforms) or social science (man-dominated) aspects of distribution.

Plants collectively are referred to as vegetation. Tansley (1939) points out that 'plants are gregarious beings because they are mostly fixed in the soil and propagate themselves largely in social masses, either from broadcast seed (or spores) or vegetatively, by means of rhizomes, runners, corms or bulbs and sometimes by new shoots ('suckers') arising from the roots. In this way they produce vegetation'. Numerous attempts have been made to classify or regionalize this vegetation but no method has been found to be perfect. This is an inevitable result of the fact that in any two areas of any given size the vegetation is unlikely to be identical. This applies at both large and small scales. Nevertheless, in order to describe and analyse one must first simplify or classify the mass of real-world data in an attempt to relate the distribution of vegetation to various factors of the environment. When Webb (1954) asked the questions 'is the classification of plant communities either possible or desirable?' it was this problem that he was highlighting.

Early workers in this field frequently related *plant communities*, the classified subsets of vegetation, to human communities, but this parallel was unfortunate because of the multiple species nature of the plant community and its less complex interdependence between individuals. The major reason that any particular area of vegetation possesses a community structure is that its species composition is limited by tolerance to that area. (Ecologically, tolerance is the excess or deficiency of an element or condition of the environment.) The detailed study of plant communities is essentially a twentieth century phenomenon initiated by Warming's (1895) treatise on 'Oecology of Plants'. From this the application of quantitative methods and statistical procedures in the analysis of structure and species distribution within the community has developed. Over time, too, the main efforts in vegetation analysis have changed scale from the extensive study of large areas of the earth's surface to intensive study of much smaller units.

In this book it is intended to introduce the geography student to the role that vegetation plays in our total spatial understanding in natural and man-altered landscapes. The study of vegetation dates back over many decades and covers many disciplines. Thus the development of historical concepts is reviewed. Because vegetation is dynamic and infinitely variable over space a chapter is devoted to succession and the difficulties of analysis of vegetation at different space scales. This is followed by two chapters outlining the standard methods of sampling and analysing vegetation to produce data suitable for geographic use.

2 Vegetation in Geography

Geography, in a general sense, is the study and analysis of the differentiation and distribution of earthly phenomena. This includes its physical features, its climates, and its products both abiotic and biotic. For much of this century, however, the only biotic products of the earth's surface which have received geographic attention have been man and his cultigens. Tivy (1971) has referred to vegetation study as the 'Cinderella' of Geography, neglected and underdeveloped despite the considerable lip-service that has been paid to its significance. Some, such as Stoddart (1965), have drawn attention to the value of vegetation in geography within the integrated holistic approach of the new 'systems' studies. Others, such as Edwards (1964), have seen biogeography as a lynch-pin around which the rest of geography could revolve, in order to contain a fragmenting discipline.

The Importance of Vegetation

None of these methodological reasons are needed to see the importance of vegetation in geography, for one of the most outstanding features of the land-surfaces of the world is the general presence of a plant-cover. The continued increase in the total human population makes the study of vegetation vitally significant since man is dependent upon plants for the very basis of his existence. Green plants provide our food either directly, or indirectly via animals; they provide clothing, housing, industrial raw materials, and over much of the world they condition the environment in which we live. The historical geographer is much concerned with the clearance of vegetation from the landscape or the transportation of plant species around the world. The geomorphologist is involved with vegetation as it affects weathering via humic acids, water infiltration via its interceptive cover, erosion via its root systems, and accretion of sediments as in sand-dunes or salt marshes. The human geographer, of course, studies vegetation within agriculture but should also know of the role of vegetation in landscape, both urban and rural. The modern development of environmental science within geography makes the study of vegetation vital within conservation, pollution, planning and environmental change. Yet it is true to say that for most of those students for whom vegetation is an important aspect of their subject, the plant-cover is no more than a frightening and unknown mass of green, shrouded in technical terms and Latin names.

The earth's plant cover varies enormously from place to place. Sometimes, as in deserts, or in the high arctic it may almost disappear; elsewhere, as in forest, vegetation is diverse and luxuriant. In areas of low relief vegetation may be the only obvious feature of the land-surface; at a macro-scale this may show up as a belt of trees picking out a small valley within grassland; at a micro-scale it may be *Odontites verna* picking out a 2–3 cm lower depression within that grassland, or *Urtica dioica* marking the site of abandoned human use no longer discernable by other surficial features. Obviously then, the geographer must take some account of natural vegetation. An increasing proportion of the earth's surface is no longer covered with natural vegetation but a secondary growth related to man's past or present interference. This should be of even greater relevance to the geographer who considers a central theme of his discipline to be man–land relationships.

Vegetation and Geographers

For the historical geographer of medieval Britain, the clearance of the woodlands is one of the most important topics; the same is true of eighteenth–nineteenth century North America. Much is written about the siting and development of settlements, the tools used and the types of soils and topography that could be cleared but little is mentioned about the type and composition of the woodlands involved. A knowledge of the trees, resulting from visits to remnants of ancient woodlands, plus the use of more recent palynological evidence and foresters information (e.g. Edlin 1966, 1973) would help considerably to add greater balance and life to the subject.

Sparks' classic (1960) geomorphological text includes mention of vegetation with reference to pollen analysis, sand dune and salt-marsh accretion, erosion rates under different vegetation types, effect on river channel cross-section, effect on chalk landscapes, and result of lack of vegetation in periglacial activity. In each case the vegetation only gets passing reference, yet whole texts in other disciplines are devoted to related vegetation–geomorphology relationships (e.g. Salisbury 1952, Ranwell 1972, Kitteridge 1948). An example of the geomorphologist's perfunctory treatment of vegetation is that, although most students appreciate the physical and climatic components in sand dune development, the only vegetation factor normally seen to be important is marram grass, *Ammophila arenaria*. In some cases the position of embryo dune grasses is overstressed whereas the vital role of the shallow rooting grasses and herbs between *Ammophila* clumps hardly gets mentioned. Similarly, the place of dune annuals and other herbs in supplying humus goes untreated. In the same way planning has become an important aspect of human and social geography yet the ideas of vegetation in planning, such as those discussed by Fairbrother (1970), attract little notice.

Vegetation and Land-use

Tansley (in Adamson 1938) remarked that a knowledge of what nature produces when she is left to herself is one of the indispensable requisites of wise exploitation. This comment emphasizes the important applied synthesis of vegetation in the landscape, which is so relevant to the geographer yet so rarely appreciated. In a broad way farmers have realized for centuries that the potential productive capacity of particular areas can be seen in the natural vegetation that is growing on them. For this reason, for instance in southern Britain, some areas have been left forested, some as heathland, and others have been turned into arable or grassland. The vast acreages of oak-hazel woodland developed over medium-heavy boulder clays is the land that has become the best extensive wheatland in the country, whereas the heather--birch landscape of the light breckland soils or the grasslands of the thin chalk are more suited to barley. Conversely, acidic *Calluna* heathland, such as is common south-west of London, tends to be too poor for the growth of crops, and the hornbeam-oak landscape of parts of Hertfordshire and south Essex is too heavy to work as arable land. Parallels can be drawn concerning the agricultural suitability of different landscapes in other countries, and it has been found that primitive agriculturalists of New Guinea and Guyana use an intimate knowledge of variations in forest vegetation in planting their gardens. If modern agriculturalists of Britain had a more intimate understanding of the vegetation on the heaviest clays, some of the severe problems of compaction when using heavy machinery might never have arisen. The lack of appreciation of the natural vegetation in East Africa at the time of the 'ground-nut debacle', or the lack of foresight illustrated when Fordlandia displaced the original forest of part of Amazonia, are further examples of where a knowledge of the local biogeography could have saved large quantities of money.

Prediction

Outside the polar, desert and high alpine lands of the world, there are few major areas that are not covered by plants to a considerable degree. Consequently the vegetation characterizes the local landscape, and because it is composed of different plants having different requirements and different ranges, it provides a valuable field of biogeographic study for interpreting landscapes and predicting the best uses to which various areas can be put. For instance, the detailed study of *machair* vegetation carried out by Dickinson *et al.* (1971) provides a framework within which the correct human use of this landscape can be better assessed. In viewing a landscape it is chiefly the vegetation which one sees. This provides an immediate and evident means of distinguishing landforms and other local features which compose the landscape. As such it is particularly useful in primary terrain analysis and in turn will be valuable to the biogeographer in land-use prediction since the vegetation is a more reliable indicator of action and interaction of local factors than direct measurement. This is because in agricultural and allied planning, as well as in vegetation growth, events of low frequency and high magnitude are as important as day to day activity.

Indicator Species and Communities

Sinker (1964) suggests that the geographer is, on the whole, more interested in vegetation than the components of that vegetation; the individual plant species. Nevertheless, in landscape appreciation the geographer should know of certain 'indicator' plants, or groups of plants, which require the presence of certain conditions (or the absence of others) and are thus valuable in demonstrating that these conditions occur where they grow. Many plants have too wide an ecological amplitide to be useful as indicators but for others their existence in a given location is an expression of actual habitat conditions obtaining locally. A useful example of this is Spence's (1957) analysis of the vegetation of Unst, Shetland, in which he shows that the occurrence of *Cardaminopsis petraea*, *Antennaria dioica*, and *Cerastium nigrescens* perfectly define the distribution of serpentine rocks. In the Breckland of East Anglia the depth of sand over chalk can quickly be appreciated by the presence, or absence, of *Deschampsia flexuosa*. Similarly M. M. Cole has shown the use of geobotany in mineral exploitation.

In order to appreciate the role of indicator plants or indicator communities the geographer must know enough about vegetation to realize that local and detailed knowledge is necessary to practise the principles involved—hence the usefulness of recourse to local floras, many of which have been recently produced or revised. Sedges, for instance, are well known for growing where there is an abundance of water in the soil, yet there are, in fact, wide variations in this respect between different species of the genus. Indeed, one species (*Carex arenaria*) is an indicator of arid conditions. Similarly Willows in general favour wet areas but *Salix caprea* may be found in quite dry soils. Moreover, members of an individual species often vary in their environmental needs according to the location within their range.

Range knowledge may be particularly important. In England the native distribution of *Fagus sylvatica* is on the chalk and soft limestones of the south, which might lead one to think of it as a calicole (*a plant of calcareous soils*). However, in Eastern Europe and in fact where it has been planted in Britain out of its native range, it will grow well wherever conditions are dry and well- to excessively–drained. Thus *Fagus* is an indicator of moisture régime and not base status.

Many geographers who appreciate the importance of higher plants as indicator species have not realized the significant role that lower taxa may play in geographical understanding even though their impact on the landscape is less. Watson (1968) provides 24 excellent habitat lists which show the value of the common British mosses and liverworts as community indicators, and Hawksworth and Rose (1970) produced evidence in the British lichen survey that lichens are one of the most delicate of all indicators of environmental conditions.

One further problem in the use of indicator species is that some species may have strains physiologically adapted to a special set of environmental conditions but lacking any evident differences in form from the norm. For instance, *Agrostis tenuis* produces strains physiologically adapted to growth on slag heaps; visually these strains are identical to those of a natural grass–land. Thus, because soil and climate may vary quite considerably over short distances, individual species of plant genera are in general unreliable as indicators. Far more reliable are groups of species or, preferably, complete assemblages—in other words the surface features of the landscape. Only species with very particular and recognized tolerance ranges should be used alone.

It is not only as gross landscape indicators that vegetation has a value. Certain groups of species may be indicative of old land-use. Harper and Sagar (1953) were able to show the indicator value of three species of *Ranunculus* in identification of old ridge and furrow. *Stellaria media* and *Ulex* spp. may often pick out a midden, or other area, of loose soil with nitrogen enrichment.

In a discussion of the role of vegetation in landscape it is worth reiterating that vegetation is not only an indicative tool of the geographer but also a formative agent. In nutrient cycling and the study of denudation rates there is a useful link-up with geomorphology. Sand-dune, lacustrine infill, salt marshes, mangrove swamps and coral reefs all illustrate the way in which the landscape is moulded by plants in a positive fashion. Humic acid weathering, joint forcing and surface water concentration are examples of negative moulding. By making use of these capacities of vegetation in the landscape the applied geographer/conservationist has an excellent tool as well as subject matter.

Vegetation and Conservation

Conservation has become an important area of study for the geographer during the last decade. The term is a wide one but it can be defined as 'taking care of the environment'. Geographers have figured widely in conservation, as books by Stamp (1969), and Warren and Goldsmith (1974), testify. In the early post-war period conservation was regarded by many as putting a ringfence around areas of Land-Use—Capability VIII (Graham 1944): those areas suitable for wildlife but not for cultivation, pasture, or woodland. Even at this level vegetation was important for finding the location of such areas of wet, rough, acid, or severely eroded, land. However, conservation was soon seen to be an active exercise involving interventionist management based upon controlled experiments over a much wider area. In Britain especially so little land remains with natural habitats untouched, and wildlife viable in the long-term without help, that environmental conservation and management has to spread far beyond the poorest land. Obviously National Parks and Nature Reserves are the prime areas of conservation but it is an ethic that must permeate the thinking of all land-use if nature is to survive as populations grow. Leopold (1966) appreciated this ethic of conservation before the second World War and O'Riordan (1971) evaluates the principles of environmental management as they appear to the geographer.

Ecosystem Studies

The biogeographer can best approach the conservation and practical management of such vegetative habitats by observing the response of the whole ecosystem to external pressures—the most obvious of which is environmental change initiated by man. Responses can be studied using many of the techniques described, from simple structural description through quantitative analysis of sample stands to, where necessary, measures of absolute abundance of species populations within a given area. The latter technique is more commonly used in rare species conservation rather than habitat conservation. *Cicerbita alpina* is presently being studied in this manner in Angus and south Aberdeenshire, Scotland. However, management at this level requires detailed knowledge of plant and environment and is more the concern of the ecologist or botanist. Furthermore, concern for a species rather than its habitat leads to fragility of that habitat and the likelihood of expensive and unrewarding control measures. Rare species can best be conserved within the framework of habitat conservation. Unfortunately, far too few of our protected, endangered and vulnerable plants in Britain are as yet within nature reserves.

An alternative approach to vegetation conservation is to combine it with detailed study of energy flow and mineral cycling within the ecosystem (Watts 1971). In this way changes to the whole system can be predicted when any one part is altered. Practically, however, despite the tremendous work carried out by the International Biological Programme, the time involved to carry out such quantitative studies makes them no more than academic, while the lack of knowledge to date, for most communities, is extreme.

For the applied geographer the empirical approach to vegetation conservation means that the desired end product must be fully appreciated and understood before management is started. For instance, biogeographers concerned with the effects of recreation on the landscape have shown that without conservation the biological system will continue to simplify as use intensifies (e.g. the major footpaths of the Lake District). Ecological principles accurately describe these effects but do not tell us at what stage management activity should intervene. Nevertheless, empirical studies of vegetation response provide hypotheses which can be experimentally tested and which can predict responses elsewhere. Regression models can extend this prediction.

Vegetation Evaluation

Much of the foregoing discussion on vegetation conservation has considered the nature reserve as the primary land-use of an area. However, in most of Europe, and in certain other areas of the world, the land surface is used intensively by man, and vegetation conservation must be included within other activities or even re-created. All living systems, however small or however altered by man, will have some interesting vegetation. Thus, in policy making, statutory bodies must attribute a value to this vegetation. Unfortunately, of the land managers sympathetic to nature conservation, more are concerned with birds and animals than the vegetation within which they live. Most ecologists feel that their researches have not gone far enough to place absolute values on ecological features: hedges, woodlands, water meadows etc. Watt (1973) has collected together some information on this subject. It seems that an ecological evaluation of any particular area is the quantitative worth a competent ecologist will put upon that particular system. In the past this worth has been relative rather than absolute; related to proximity, to research and to previous work.

The techniques of evaluation best employed today are to relate the various examples of particular habitat types; forest, grassland, water, on a scale of values that are entirely ecological. These might be species diversity, number of vegetation layers, areal extent, definition of boundary, age and rarity of habitat. If the vegetation of the landscape as a whole, within a particular area, is to be evaluated, there is a difficulty in comparing such features as ponds, hedges, and old fields within a single scale. Similarly there is a problem in giving values to small woodlands as against few large forests of equal area. Ecologically the fewer large forests would be more valuable because there would be less side effects, but potentially they would be at greater risk in a changing environment because of their smaller numbers.

O'Connor (1973) suggests that it is possible to devise a number of indices of ecological value with numerical notation ascribed to habitat features. These may be additive or multiplicative and could provide a concise and objective statement about any given site. Evaluations for larger areas can be derived from a set of habitat evaluations by weighting particular habitat-types according to their areal importance. Ecologists and biogeographers have a responsibility to provide input to planning processes and, however tentative such schemes may sound, they can only be improved by successive approximation (Poore 1955) beginning with this kind of technique. University College, London have published a series of Discussion Papers in Conservation showing some practical examples of these theories and techniques which are highly relevant to the applied geographer and biogeographer. The extreme aspect of this type of work has been termed creative conservation, made necessary by the progressive increase in scale of landscape modification. In most areas it is relatively simple to predict the vegetation a given area, such as a motorway verge or exhausted gravel pit, will support, but much must be known about theories such as succession if the required vegetation is to be recreated. Because agronomists and horticulturalists know of the relationships between grass species and edaphic conditions, most of our motorway verges become artificial meadows. The theories and techniques of vegetation analysis could well be put to use in conserving natural communities within these areas.

Pollution

One area in which detailed knowledge of the internal functioning of the ecosystem is needed is that of the relationship of environmental change to pollution. The direct effects of a pollutant, such as a herbicide, or gaseous emission, can be measured in terms of reduction in diversity of the vegetation complex; but indirect effects, such as the toxicity of residues in the soil, are much less clearly appreciated without recourse to energy flow and nutrient cycling. Many of the herbicides that are marketed as non-poisonous have been shown to alter the composition of soil microfauna and microflora if repeated applications are made. Where experiments have been carried out using permanent quadrats it has been shown that these soil changes affect the quantities and qualities of species other than those at which the herbicide was directed.

A common technical problem with the use of herbicides is chemical drifting. Mellanby (1967) provides evidence of many areas of surrounding vegetation that have been damaged by misuse of herbicides in agriculture. Often this damage is selective and only obvious when it is well-advanced, unless it is monitored by using one or other of the sampling techniques available. It is even arguable whether it is worthwhile economically to remove all weeds within agricultural land. This is an area of research in which both biogeographers and agricultural geographers could become involved. Many agricultural landscapes might be aesthetically improved without loss of crop yield if routine herbicide spraying were discouraged. This is particularly true as the cost of labour and chemicals increases more rapidly than the value of the crop. Only a quantitative analysis of the vegetation can ensure an economic advantage to the farmer as well as an increase in vegetation diversity.

Road Networks

Roadside verges are areas particularly susceptible to herbicide pollution in many counties, yet they are increasingly important sanctuaries for natural vegetation and wildlife, even though they are entirely artificial and man-made habitats kept at one successional stage and thereby maintaining a wide diversity of plant species. Highway authorities have a duty to maintain verges in order to stop vegetation from encroaching on roads and to improve visibility. Herbicides such as MCPA and 2.4-D are still used in some areas to do this. These effects can best be assessed with a sampling and analysis programme (Yemm and Willis 1962).

Road salting is another form of pollution that alters verge vegetation, producing salt values in the soil comparable to those found near the seaward limits of plant growth in coastal sand dunes. Circumstantial evidence of the role of salt pollution in road verge vegetation can be obtained from comparative analyses of salted and unsalted roads, correlation of amount of damage with amount of salt used, death of trees and shrubs near salt dumps, and comparison of the floras of coastal cliffs and salt-affected verges. Ranwell *et al.* (1973) describe techniques suitable for analysis of salt effects on road verges and suggest suitable species for planting along salted roads.

Air pollutants, too, may affect roadside verge vegetation and also vegetation elsewhere. On the M45 the soil of the central reservation was found to contain 500ppm lead (Ranwell *et al.* 1973) from exhaust gases. Hawksworth and Rose (1970) showed that epiphytic lichen communities act as biometers; living measures of the levels of pollutants in the environment. They recognize, via sampling and analysis of the lichen flora of England and Wales, a series of ten noda which form a continuum developed along an air pollution gradient. Using these noda ten zones of atmospheric purity can be defined according to the species groups found on deciduous trees with rough bark. Trees rich in foliose lichens and sensitive species such as *Teloschistes flavicans* only occur where air is pure; above $150\mu g/m^3$ SO_2 in the atmosphere only *Lecanora conizeaoides* remains. Evidence of this sort is necessary to present to the relevant authorities to bring about a political solution to the dangers of pollution.

Vegetation Change and Man

Dansereau (1947) has been able to scale man's impact upon the vegetated landscape into successive phases which illustrate an increasing measure of his control. At the lowest level is the development of a human society which lives by gathering vegetable material. Certain tribes of Amazonia still practice civilization at this level but cause no long-term effects on the ecosystem. At the most they temporarily deplete seed supply or modify vegetation structure of the landscape but so, too, do other primates. Hunting, too, is a simple form of civilization but this can induce sensible modification of the biotic environment. It has been suggested that the savannas and prairies may well owe their extent, if not their origin, to man's use of fire as a tool in hunting (Hills and Randall 1968). Biogeographic research may eventually solve the savanna origin dilemma via vegetation studies. At the higher level of development herding is practised. Thus usually causes a vegetation change on a longer timescale. For instance the grazed lands of montane Europe are not natural grasslands but revert to coniferous woodlands when grazing is discontinued. Just as the agricultural geographer is concerned with the economics of differing domesticate species numbers per acre, so too can the effects upon the species composition of the pasture or the return to woodland be analysed. The effect of annual fire is frequently added to that of grazing. In the Rupununi savannas of Guyana, when the grasses die down and leave the cattle without food, man induces new growth by fire. Regularly fired areas show a different species composition to those further away from villages, where fire is less frequent, despite the fact that all the savanna is pyrophyllous.

Vegetation Analysis

There are still some herding communities who use natural pastures. However, even in these conditions of extensive grazing the most palatable species will be eaten out and eventually range composition will change. Unless the geographer appreciates the effects of grazing upon range vegetation he will not be able to predict future land-use.

Agriculture

Agricultural land-use introduces domestication of plants, an area in which the biogeographer or agricultural geographer should have some understanding of plant genetics as well as vegetation ecology. Agriculture is a war between cultigens and natural vegetative regrowth. A knowledge of succession of vegetation communities allows agricultural man to take advantage of those stages which are most advantageous for his purposes in different parts of the world. Hence the cereals of North America are so successful because they are grown in an area where soil and climate are preadapted to grassland. Industrialization is a form of civilization which substitutes natural ecosystems and leads eventually to the need for conservation and environmental change monitoring already discussed. The effects of industrialization on vegetation are manifold, though usually indirect via habitat change. For instance, the development of reservoirs for hydro-electric power, or water-table reduction via heavy industrial water-use, can create water-levels or fluctuations out of step with local rainfall and drainage patterns. This in turn can affect the local vegetation. Excessive extraction of water in dune areas has been shown to eradicate dune slack communities containing *Juncus articulatus*, *Salix repens* and *Erica cinerea*. An enrichment of ground-water in dune slacks via industrial effluent can cause the local extinction of species such as *Parnassia palustris* whereas the flooding of dune slacks by raising the water table, as has occurred in the Wassenaar Dunes, Holland, will also remove the dune slack vegetation, and replace it with open water. Factory dust and fumes may similarly have an effect upon the local vegetation. It may not kill the plants directly but lime dust from a cement works has been seen to reduce the rate of growth of certain species, resulting in competitive disadvantage. The final effects of man's intrusion upon natural ecosystems is urbanization; that state in which the wild vegetation has been totally replaced. As Dansereau (1947) points out, these stages do not merely tend to replace one another but may be concurrent. Many of the techniques of sampling and analysis discussed in later chapters could be used to monitor these changes.

Vegetation Dissemination

One of the major themes running through human and historical geography has been the tracing of innovation dissemination through space. A similar aspect of biogeography is the study of movements of plant species as they are affected by man. Man has had similar effects upon vegetation distribution over short periods of time as plate tectonics, glaciations, marine invasions, and ocean currents have had over geological time.

Man assists the dissemination of vegetation both deliberately and accidentally, sometimes being conditioned himself by physical forces (such as the Polynesians in Oceania) and sometimes operating against them (such as Europeans sailing to the West Indies). Randall (1968) investigated the original distribution of 100 plant species of the Barbados coastal vegetation and found that 23 had been introduced since 1627 (Table I). Some of the earliest recorded introductions were grasses: *Eleusine indica* is a good pasture species and was most likely brought in deliberately, whereas *Egragrostis ciliaris* is a very poor pasture type and its introduction may well have been accidental. *Swietenia mahagoni* and *Casuarina equisitifolia* were introduced as windbreak trees and *Delonix regia* and *Terminalia catappa* as ornamentals. Old manuscripts and floras are often the best source of information of this sort. For instance Hughes (1750) describes the import to Barbados of *Cyperus rotundus*: 'It was brought here in a pot of flowers sent from England to Mr. Lillington in St. Thomas parish: from thence it hath been more or less unluckily propagated throughout the whole island.' During the same period since 1627 man has also altered the coastal vegetation in a negative manner: drainage has eradicated two mangrove species: *Laguncularia racemosa* and *Avicennia germinans*.

TABLE I.

The original distribution of the Barbados coastal vegetation

Original Distribution	Number of species
Pantropical	20
Panamerican tropics	14
Caribbean	12
Caribbean and central America	9
Caribbean, central and South America	7
Tropical Atlantic	6
Caribbean and South America	5
Caribbean, central and North America	2
Endemic	2
Introduced after 1627	23
	100

Vegetation Analysis

There are several techniques available for attempting to understand man's role in vegetation dissemination. Randall (1972) has reviewed some of these for the sweet potato. There are the archival records already mentioned. These include ships' logs, lexicography, explorers' note books and missionary documentation. On occasion experiments have been conducted such as Thor Heyerdahl's (1950) *Kon Tiki* expedition which proved the feasibility of westward travel across the Pacific. Carter (1953) concludes from this that:

Had cross–Pacific voyages been long and arduous, the sweet potato tubers would either have spoiled or been eaten. Neither was the case, and it is clearly inferable that man made deliberate and relatively easy voyages across the great expanses of the Pacific.

Genetics, too, can play an important part in understanding man's role in plant distribution. Yen's (1963) work on the varietal population of the sweet potato, *Ipomoea batatas*, has shown that it is a single species throughout those regions where it is cultivated, since there are continuous ranges in individual morphological characters but consistent chromosome numbers. If the Oceanic sweet potato were to have been the result of a single transfer from the Americas, then there would have been a simple east–west decrese in variation. However, this does not appear to be the case: four Asiatic and five Pacific characters exceed the corresponding ranges in New Guinea, but only two of these chaacters are common to both Asiatic and Pacific populations. This would seem to suggest three separate introductions over a considerable period of history—possibly innovation waves of pre–historic American introduction, sixteenth-century Spanish introduction and overland introduction from Asia.

An analysis of floral records at different times in the recent past is now possible, through the publication of second or further editions of county or provincial floras in Europe and North America and through distribution atlases such as Perring and Walters (1962). Such publications illustrate innovation and spread of plants either concentrically from their point of import or linearly via railways and roads. The spread of *Erigeron canadensis* across North America is one of the best documented examples.

A Summary

Dansereau (1957) sums up man's effect upon the vegetated landscape by categorizing nine elements that are affected. Changing conditions, such as in forest areas which have been lumbered but not deforested, result in the spread of indigenous elements like *Crataegus* spp. within the woodlands of the St. Lawrence valley. Sporadic elements occur occasionally in scattered localities but never quite establish themselves as permanent members of vegetation community: *Papaver rhoeas* in North America is an example.

Exotic elements planted and protected by man are one of the most significant aspects of many vegetated landscapes, yet they do not maintain themselves spontaneously. The vast majority of crop plants are of this type but so, too, are the *Eucalyptus* groves of Brazil and Israel, planted to assist in swampland drainage, or the *Picea abies* forests of Canada. Less important in their spatial impact are the exotic elements naturalized within either the home or the urban landscape. House plants, such as *Selaginella caulescens* in bottle gardens, are an example of the former, while many city treescapes exemplify the latter: *Galinsoga ciliata* in Montreal or *Sorbus intermedia* in Norwich.

Exotic elements may naturalize in waste places and produce vegetation communities which are closely related to the history of the area. *Opuntia ficus-indica* on Mediterranean wasteland results from the early Spanish and Italian exploration of Central America; the presence of *Phleum pratense* in the forests of Canada is evidence of an old lumber camp. Exotic elements may only naturalize in cultivated fields such as *Leucanthemem vulgare* in the American prairies, or in disturbed habitats like *Poa annua* on tropical forest paths. The most extreme examples of man's effect upon the vegetated landscape is where exotic elements he has intoduced have naturalized in the primeval habitats of the new country. Most introductions cannot compete ultimately within the indigeneous flora unless there has been human interference. However, *Butomus umbellatus* is slowly displacing the riparian vegetation of the St. Lawrence while *Elodea canadensis* is widespread and once was abundant in slow–flowing fresh water in Europe after having been introduced in the mid–nineteenth century. On sea cliffs in Cornwall, Devon, Suffolk, and the Channel Islands *Carpobrotus edulis*, a South African mesembryanthemum, hangs in clusters and completely displaces the natural vegetation. It also occurs widely over the dunes of Mediterranean Israel. Occasionally one finds a man–introduced species that has naturalized in climax woodland: in the *Acer saccharinum* forests of Quebec, *Epipactis helleborina* has spread almost certainly from the gardens of early settlers. In such situations the impact of the man–introduced species upon the vegetation as a whole is usually small.

Man's use of fire, whether for hunting and farming, as discussed earlier, or for war or by accident, is one of his most significant methods of altering vegetation and as such is important to the geographer. Batchelder and Hirt (1966) suggest that ninety eight per cent of vegetation fires are caused by man. Many geographers assume that non–

cultivated land is covered with 'natural' vegetation whereas an analysis of the composition of many communities will show a high incidence of pyrophytes; plants that resist burning, or are stimulated to sprout after fire. The vast majority of these owe their present extent to man (Hills and Randall 1968), though of course there are natural pyrophyte communities owing their existence to lightning strikes or vulcaninity, such as the *Imperata cylindrica* grassland of the Philippines.

Many heathlands are held static by fire since the woody roots of *Vaccinium* spp. and *Calluna vulgaris* will soon sprout vigorously after a controlled burn provided that the peat itself does not catch fire. In the tropics, palms and palmettoes are indicative of frequent fire and considerable biogeographic work has been carried out in the Everglades National Park to assess the role that man–induced fire plays in creating the southern Florida landscape. Grasses are especially well adapted to fire, the genus *Imperata* almost always being indicative of regular burning in the savannas. Pines are the major pyrophylous trees, covering vast areas where periodic fires occur in Canada (*Pinus banksiana*), the Landes of France (*Pinus pinaster*), California (*Pinus contorta*), the Philippines (*Pinus insularis*), and middle America (*Pinus caribaea*). A knowledge of these or similar species and the role they play in the landscape is of great use to the geographer in understanding the effect of man on the environment.

3 History and Development of Vegetation Study

The Themes

Field studies of vegetation began in the early nineteenth century with the work of Alexander von Humboldt (1805) in plant geography. This was a study of the spatial distribution of taxa and their evolutionary relationships, and has become the classic biogeography of the natural sciences concerning dispersal (Ridley 1930), succession (Clements 1928, Whittaker 1953), taxonomy (van Steenis 1969) and genetics (Yen 1963). Over time, certainly by the end of the nineteenth century, the study of communities and community composition widened the field to include physiological ecology and plant sociology (Schimper 1898. During the twentieth century further research has added population ecology (Boughey 1968), genecology (McMillan 1960), ecosystems (Odum 1959), and palaeoecology (Faegri and Iverson 1964).

The various areas of specialization within vegetation study have, by and large, developed independently, some primarily in Europe and others in Britain and North America. Similarly geographers have concerned themselves with only parts of the subject matter: at first only plant geography and palaeoecology and, more recently, plant sociology in Europe and ecosystem studies in North America and Britain. The blanket term 'plant ecology' has often been used to cover vegetation study. Ecology was a term coined by the German zoologist Haeckel in 1866 with reference to the teaching ($\lambda o \gamma o \varsigma$) about the household ($\dot{o} \kappa o \varsigma$) affairs of organisms. In the English speaking world ecology has come to include the study of groups, populations, or individuals in relation to their environment and covers description, mapping, measurement, and experimentation. In Europe the term has tended to exclude all plant sociology, plant geography, and historical studies.

Thus the approaches that have developed towards vegetation study are very widespread. The early literature of European travellers to unknown parts often concentrated on vegetation because it was such an important feature of the landscape. However, detailed description is cumbersome and this century has seen attempts to simplify via codification of data and tabulation of results. Clements (1928), Du Rietz (1921), Braun–Blanquet (1965), and Raunkiaer (1934) are the best known workers of this type. Unfortunately, as their work progressed, it became more specialist and less relevant to students of vegetation because of its dry presentation. In recent years there has been a marked reawakening of interest in large scale spatial description of vegetation as a result of the IBP (International Biological Programme). This can best be seen in Peterken (1967). Such general qualitative description of vegetation is a vital prerequisite to any further study.

Phytosociology

In Europe this has meant, above all, plant sociology. Braun–Blanquet (1965) developed a system of phytosociology based on a hierarchy of systematic units which parallels the Linnean system of taxonomy. Using this system, Braun–Blanquet hoped to find the functional and causal reasons for spatial distribution of plant-types.

American and English researchers did not accept the Braun–Blanquet system. Although they used floristic data their classification was based upon dominant species or species groups rather than on differential species. The resultant communities were usually larger in size and more heterogeneous in physical environment simply because dominants have wider ranges of tolerance. The main trends come from Clements *et al.* (1929) in America and Tansley (1939) in Britain.

Plant Ecology

Plant-habitat relationships and the classification of vegetation by environment was one of the major themes that developed from Warmings' (1909) original work. Today this is often seen in attempted syntheses of floristics and environment such as that of Loucks (1962), Daubenmire (1968), and Odum (1959). A primary outcome of this synthetic treatment has been the need to confirm hypotheses generated in the field by laboratory experimentation. This is one area where geography students have rarely ventured, because of lack of facilities.

One of the reasons that the European hierarchic scheme did not appeal to the English-speaking world was its static nature which did not fit with the ideas of succession advocated by Cowles (1899) and Clements (1928) and supported by Tansley (1929). These ideas developed in the more continously distributed and less man–altered landscapes of the New World, where a succession towards a regional climatic climax seemed to fit. Whittaker (1953) has considerably modified the initial ideas of succession without removing any of the dynamism they contained. Dynamism within vegetation is also basic to the continuum concept of the Wisconsin school (Curtis 1959, Beals 1960) and the synecology of Daubenmire (1968).

Objective Methods

In the English–speaking world a great desire for objectivity in community definition resulted in the European techniques being rejected over much of this century. In his individualistic concept Gleason (1939) made the observation that no two communities are alike. In the forests and grasslands of North America self-evident communities with absolute boundaries, such as Braun–Blanquet sampled in the Alps, do not exist. Thus the Wisconsin school (Curtis 1959, Cottam 1949, *et al.*) studied communities through sampling individual populations randomly or systematically along selected distribution gradients. The result of this work was the realization of the continuity of distribution of both populations and communities. In Brtiain the desire for objectivity resulted in the quantitative investigations of much smaller vegetation patterns (Greig-Smith 1957, Kershaw 1964, Lambert and Dale 1964). Pielou's (1969) recent book is a useful analysis of modern mathematical interests. Because the mathematicians in vegetation study have emphasized continuity the ideas of classification were rejected in favour of ordination (Bray and Curtis 1957). However, the most modern techniques (Wishart 1969, Goodall 1970) make mathematical classification of continuous data eminently feasible.

The Ecosystem

One of the most successful trends within vegetation studies in the Post War period has been the holistic study within the ecosystem framework involving the community plus its habitat (Tansley 1935). Obviously many authors have included several of the themes discussed. Similarly many people have thought along the lines of the ecosystem (Forbes 1925, Major 1969). But it was not until Tansley produced the term in Britain and Sukachev (1945) and Odum (1959) used the ideas in Russia and the USA respectively, that the functional stability and interaction of the ecosystem was fully appreciated. The great advantage of the ecosystem is that it provides a theoretic basis for geographers, biologists, and others to study their own aspect: vegetation, soil, plants, animals, climate, man; yet enables a synthesis of the findings of all. Ecosystems have no limitations in space or time and thus do not replace the concepts of vegetation discussed earlier; but they do have the advantage of fitting vegetation studies into a totality of understanding that gives the whole so much more life. Kellman (1975) rightly criticises the ecosystem in terms of difficulties of mensuration and application to productivity, but in theoretical terms it is so far unsurpassed.

Vegetation Analysis

The Hypotheses Organismic and Individualistic

The themes that have been outlined above have been based upon a series of hypotheses that have coloured vegetation studies throughout its history. These are therefore worth examining in greater detail. Two hypotheses have led to more misunderstanding than all others: the organismic and individualistic. Initial ideas of Clements (1928), and Cowles (1899), thought successional development of a plant community akin to the birth, ageing, and death of an organism. Just as an organism reproduces, so too a climax community can reproduce by repeating its stages. The limitations of this analogy are that death results not from senility but through environmental change and competition. Similarly, growing and maturing are replacement, rather than developmental, activities. Furthermore, reproduction can only occur in like environments. Tansley (1929) criticised the pure organismic hypothesis and replaced it with the idea of the quasi-organism—emphasizing that because the plant community acts as a unit, it should be studied as a unit. It was from this hypothesis that the ecosystem developed.

Followers of the organismic hypothesis assumed that the discrete structure of the vegetation they studied was factual; often it was the result of the classification process they used (Philips 1935). Thus classification came to be looked upon as a method only suitable for those who 'believed' in discontinuous data. By the 1920s the idea was evolving that vegetation was in fact continuous over space and could not be classified into discrete units. Gleason (1917, 1939) pointed out that species had individuality and were distributed in relation to the total range of environmental factors, including other species, and thus no community of species with measurable boundaries can exist. This idea was opposed by traditionalists like Clements, Weaver, and Hanson (1929), but eventually gained favour with Cain (1947). Whittaker (1967), the Wisconsin school of ecologists (Brown and Curtis 1952), and workers in other countries (Dagnelie 1965, Goodall 1953, 1954). Although the individualistic hypothesis is absolutely correct, when viewed on a relative basis, similarities and differences are distinguishable. This problem applies to taxonomy as well as to vegetation study but has remained a stumbling block for ecologists until recently and has had important consequences for the advance of the subject.

Whereas Gleason and Clements were arguing for a continuity of change in time (succession) and Gleason was equally adamant about space, Daubenmire (1966) distinguishes between rate of change and direction of change in time and space. Braun–Blanquet, with his emphasis on hierarchy, tends to paint a very static picture of the real world. As spatial coverage of his technique increases, so it is found that the floristically defined classes with characteristic species cannot be extended to the same worldwide application as the Linnaean system in taxonomy. This is because the characteristic species change in their ecological and sociological relationships in different parts of their range.

A Synthesis

Variations in the physical and organic parameters of the environment cause particular reactions in the vegetation cover and differences in both the relative abundance and the spatial relationships of plant species. Therefore the vegetation cover can best be though of as a mosaic of internally organized, but interdependent, plant–environment systems. It is the vegetation within these integrated systems that are the 'communities': assemblages of species which form relatively distinct and separate units. Each community so defined has its own structural and floristic unity and displays a marked organization not only between the plants themselves but also between plant and environment. The plants themselves form an integral part of their own habitat and modify it over time. The functional processes of the community are expressed by the formal structure and areal extent.

No plant community can be a self-contained unit either floristically or ecologically since there is a continuous variation in the plant environment complex. But the concept of the community, as it is frequently used in vegetation study, is based on the premise that it represents a real-world ecological unit as well as a conceptual and floristic abstraction. Poore (1956) thus defines any difference in species content between two communities as an effect of, or a factor associated with, an alteration of one or more environmental variables. Chance, though, may play a part, especially in open communities.

The classification of the vegetation cover into types, or the ordination of parts in relation to the whole, are both abstractions of the real-world that explain different facets of the community complex. As has been said they need not be related to the vegetation pattern. However, problems have arisen through confusion between the real-world community and the abstract community-type. Poore's (1955) use of the terms *noda* for the abstract and *community* for the concrete at Breadalbane, Perthshire is one answer to this difficulty. Terminology has, in fact, been one of the greatest stumbling blocks to a full understanding of both hypothesis and method. The 'association–formation' confusion was a major cause of disagreement both within and between the various national schools of vegetation study. This has been well reviewed by Shimwell (1971).

4 Vegetation Change in Time and Space

Succession

The continued existence of bare ground on the surface of the earth is a rare phenomenon associated with some extremes of climatic or edaphic condition. Normally the area is rapidly colonized by a series of plant species which will, in turn, modify various aspects of the environment. The result of this modification is usually an amelioration of the habitat which allows further species to become established. This sequence is called succession, a concept initiated by Warming and Cowles and described in greatest detail by Clements (1904). Tansley (1929) defined succession as 'the gradual change which occurs in vegetation of a given area of the earth's surface on which one population succeeds another'. Clements saw succession as having six components:

  (i)   nuduation—the exposure of a substrate
  (ii) migration—the arrival of disseminules (*seeds, inflorescenses* etc.)
  (iii) excesis—germination, growth and reproduction
  (iv) competition—resulting in species replacement over time
  (v)   reaction—habitat change through species development
  (vi) stabilization—the attainment of climax.

Clements used the word 'sere' to describe the developmental stages through which the vegetation passed until it reaches climax or equilibrium with the geology and climate of an area. Thus one can investigate a hydrosere as open water is colonized by submerged and free-floating aquatics, then rooted floating–leaved aquatics, reed, marsh or fen, carr and eventually woodland; as around the Norfolk Broads. Similarly the colonization of bare rock is called a lithosere, sand–dunes a psammosere, and salt–marshes or mangrove swamps a halosere.

Effect on Environment

The effect that vegetation has on the environment in which it is growing is the outstanding feature of succession and constitutes the main mechanism of change which allows succeeding disseminules to grow. An obvious example of this activity is found in the formation of sand–dunes and their subsequent change to woodland or grassland. Growth from a seed or rhizome of *Ammophila arenaria* results in a disturbance of the environment at the point of projection of its aerial parts. Increased impedence to the wind causes sand particles to be deposited and the plant to be buried. However, *Ammophila* has a remarkable facility for upward growth; through up to one metre of sand per annum such that as the sand accumulates so the plant grows through it and more sand accumulates. Over a period of years a progressively larger dune is formed which is then colonized on its leeward side by other plants and grasses such as *Festuca rubra*, which stabilize the sand surface. Eventually the mobility of the sand is decreased and a closed cover of grasses and herbs develops which exclude *Ammophila*. In areas of siliceous sand this may eventually be succeeded by *Pteridium aquilinum*, *Calluna vulgaris*, and *Betula pubescens*, as the sandy soil becomes less alkaline and more enriched with organic matter. Similar examples can be drawn from the role of successive plants on the salt marsh: initiation with *Salicornia* spp. or *Spartina anglica* followed by *Puccinellia maritima*, *Aster tripolium*, *Limonium vulgare*, *Armeria maritima* and *Juncus maritimus* as the ground level increases and tidal inundation decreases.

These examples of the reaction stage are obvious and are a vital element of both dynamic geomorphology as well as biogeography but there are other forms of reaction which are less obvious. Clements divided succession into two kinds: primary succession, initiated on bare land or in water, where reaction is clearly seen; and secondary succession initiated by environmental upset which disturbs a previously operating successional sequence. The firing or rabbit erosion of sand dunes, the grazing of grassland or the felling of trees in woodland are examples of initiations of secondary succession where reaction is obscure and excesis and competition much more obvious.

Whether succession will only develop progressively as Phillips (1934) suggests, 'all examples of apparent retrogression are explicable in terms of some disturbing agency', has been discussed by many authors (Tansley 1916, Cooper 1926). For the geographer the most important fact is that succession is a dynamic process involving any directional change in vegetation, due either to intrinsic properties of plants or to changing environmental factors.

Time in Succession

A great deal of literature describes succession in detail in various parts of the world but it is rare for this succession to be related to an actual or even approximate time-scale. The time-scale in primary succession may involve hundreds or even thousands of years. This depends at what stage stabilization is thought to occur. On such a long time-scale, evolutionary change in adaptation and speciation may take place. One important example of this is the equatorial region where, in places, geomorphic nudation has not occurred since angiosperms evolved. Therefore less specialized plant life may have occupied the same sites that today contain highly evolved rain forest. In many situations numerous secondary successions can occur within the course of a primary succession, while primary succession remains an open-ended activity.

Succession may be studied in one of two ways: repeated observation of the same area or side by side comparison. The former is the more reliable since no inference is involved but of course the timelength of study is greater. Studies of the same area can be based on permanent quadrats, exclosure quadrats, air photographs, historical records and evidence of change within the present community. Side by side comparison is the replacement of this time scale element with a space scale which, it is assumed, represents the time-scale. The outstanding example of dated successional sequences are those of salt marshes, sand dunes and glacial moraines.

The geomorphic, pedologic and botanical work of Cooper (1937), Field (1947), Lawrence (1958), and Crocker and Major (1955) have given a clear picture of the recession of the glacier at Glacier Bay, Alaska, and its subsequent successional stages of vegetation. The complete succession from bare glacial debris to mature spruce forest takes approximately 250 years. During this time the major reaction of the vegetation is to build up soil nitrogen, and concurrently to reduce the soil pH. As soil organic matter increases, so soil structure develops and 'crumbs' replace the amorphous glacial detritus. However the details known are only correlations, not cause-effect factors; to understand fully the mechanism of succession, autecological studies of each species involved would be required.

Salt marshes have been studied in a similar fashion. There are two methods by which their age may be estimated. One is to use old maps but as a rule little confidence can be placed in them for this purpose. Chapman (1960) records that this method has been helpful at Holme, Norfolk where the closed salt marsh has developed since 1858. The second involves using sedimentation rates in relation to depth of marsh mud and maximum tidal range. In such a case it is necessary to assume stability of sea level. Chapman (1938) worked on the Scolt marshes where he discovered that it may take about 200 years for an open marsh to reach the *Juncus maritimus* stage. The components of this succession are as follows:

Dominant species	Vertical range	Accretion rate per century	Time to accumulate depth of silt
Salicornia spp.	40cm	68cm	58 years
Aster tripolium	65cm	98cm	20
Aster tripolium + general Salt Marsh community	82cm	90cm	28
Limonium vulgare	95cm	80cm	15
Armeria maritima	112cm	36cm	5
Plantago maritima	125cm	42cm	29
		Total	201

The original work by Cowles (1899) on the sand dune vegetation of Lake Michigan was a classic contribution to the concept of plant succession despite its lack of chronology. However Olson (1958) has re-examined the successional stages of these dunes in relation to age and soil development. The dunes developed on higher 'raised beaches' dating from early Post-glacial higher lake levels up to 20m above present Lake Michigan. The oldest dunes have been radio-carbon dated to about 12 000 B.P. More recent dunes related to present lake levels were dated by tree-ring counts of *Pinus banksiana*, an early colonist. The leaching of calcium carbonate from the sand is very rapid, and the upper 10cm became devoid of $CaCO_3$ within a few hundred years. There is a slow build up of organic carbon and soil nitrogen so that by 1000 years *Quercus tinctoria* enters the succession. From then on succeeding generations have little effect upon the soil properties and an equilibrium condition results: what Cowles called the climax state. Similarly, Randall (1973) was able to use change in $CaCo_3$ content of dune soil in the Monach Isles, Outer Hebrides to assess the build up of dune grassland since secondary succession was initiated after vegetation denudation around 1810. Odum (1960) concluded that secondary succession on an abandoned field in South Carolina may involve a series of equilibrium stages associated with the prevalence of major life-forms (groups of plants with morphological features in common) rather than a continuous change with species changes as usually occurs in primary succession.

Mueller-Dombois (1965) considers that logging and fire involve a less complete sequence of life-forms in secondary succession than other forms of disturbance. Yet the recognition of stages in secondary succession is of great importance in the evaluation of a regional vegetation cover since different plant communities within the same floristic region may well result from secondary succession. Islands have a peculiar aspect of secondary succession associated with newly evolving floristic patterns caused by introduction of new species by man. This non-reversible floristic pattern rarely occurs in saturated continental areas where ecosystems are less fragile.

Monoclimax

In almost all cases vegetation successions develop to a stage—climax—in which plant life and environment are in virtual equilibrium. There have been many definitions of climax; some reflecting the monoclimax concept and other the polyclimax. The monoclimax theory was developed by Clements (1936): for any given climatic zone, vegetation communities would change until only one type of vegetation climax was present, despite the variety of initial derivation. This resulting climatic climax formation would differ in details of local species composition but would be essentially so similar that its relation would not be in doubt. An example would be the boreal coniferous forest formation of Canada or Northern Eurasia. Nevertheless, at any stage the succession towards climatic climax can be stopped either temporarily or permanently by 'arresting factors' of geomorphic, edaphic or biotic nature; this will give rise to subclimaxes in which vegetation is held stable by non-climatic controls. If the arresting factor is removed then subseral succession towards the climatic climax will recur. When man or his activities are the arresting factors one has plagioseres leading to a plagioclimax—such as the heather moorlands of Scotland. In fact the bulk of the world's vegetation in the twentieth century is plagioclimax and very little climatic climax occurs. The latter term, if used, should only be applied to that vegetation which retains its essential stability for a longer time than the lifespan of the component organisms. The delimitation of climatic climax formations has also a spatial problem because variation in local environmental situations may result in stable 'preclimax' (relict of former warmer, drier times) or 'postclimax' (relict of former cooler, moister conditions) vegetation. 'Disclimaxes' (where human interference is repeated and regular) such as *Opuntia vulgaris* communities in the Northern Territories of Australia also may occur. The overuse of this type of terminology has been criticized by many students of vegetation since many special terms merely complicate the issue.

Polyclimax

Instead the 'polyclimax' theory may be invoked, which does not assume that only one climax formation is found in a specific region

(Gleason 1939). This concept recognizes that vegetation formations may be influenced equally by climate, geomorphology, soil, the sea, man, and so on. Tansley (1935) points out that climatic climax is thus not ruled out but put in perspective. Odum (1959) sees the climax of the polyclimax theory as a stable community in a successional series, self-perpetuating and in equilibrium with the physical conditions. In fact the real difference between the two concepts is one of time-scale. Firstly, it is unlikely in many parts of the extratropical world that climate remains stable enough for long enough to allow a stable climatic climax to develop. Secondly, the polyclimax theory does not allow for a sufficient length of time for climate to become the overall controlling factor of vegetation since it is working on a much shorter time-scale.

Stability

Stability is of course a relative word and, in the life span of many insects, weed communities would be climaxes. When one thinks of the vast fluctuations of the Pleistocene then one must realize that the whole climax concept is in fact abstract rather than concrete, similar in difficulty to the geomorphic time-scale study described by Schumm and Lichty (1965). Clements' monoclimax was to ecology what Davis' (1909) cycle was to geomorphology or Marbut's (1927) zonal system to pedology. It is important to realize that the stable climatic climax theory was based upon studies of mid-latitude vegetation. Richards (1952) shows that in the tropics there are widespread stable vegetation types determined by soil conditions such as those around Moraballi Creek. These may well have developed over many thousands of years of minimal climatic variation. Although difficult to use in practice the concept of a 'biotic potential' (Watts 1971) may be more convincing. This hypothesis considers an ever-changing mosaic of environment towards which an ever-changing pattern of vegetation is striving to reach equilibrium.

Oscillation

Despite the concentration of the literature on unidirectional progression in vegetation communities towards a static end point, there are some cases in which vegetation is seen to oscillate either in time or space. This situation is probably much more common than has been reported but it is most frequently seen in the hummock and hollow cycles of peat (Godwin and Conway 1939) on raised bog, the tundra polygon cycle (Billings and Mooney 1959), and the heather/bracken ageing cycle (Watt 1955). Grasslands, too, appear to have pioneer, building, mature, and degenerate phases of a cyclical nature according to Watt (1947) as do some mature forests over a much greater time-scale.

There is often a correlation between cyclic geomorphic activity and cyclic vegetation. During peat accretion on a raised bog *Sphagnum cuspidatum* invades pools of water. As the moss grows out of the water, it is replaced in turn by *S. pulchrum* and *S. papillosum*. With the advent of the latter species the pool disappears and a low hummock develops which is colonized by *Calluna vulgaris* and other woody perennials. Eventually *Cladonia arbuscula* or in some areas *Leucobryum glaucum* develop on the heather which ages and dies. The hummock then parti-ally erodes and, being out of phase with adjacent hummocks, becomes the pool in which sphagnum mosses recur. In this way the bog becomes higher. The pioneer, building, and mature phases of the alpine tundra of Wyoming is very similar with *Carex aquatilis*, *Carex* and *Sedum rho-danthum*, and *Geum turbinatum* and *Polygonum viviparium* replacing each other in turn. Biologically, it is interesting to discover the initiat-ing factor that triggers each phase. Microclimatic variation is certainly present but it seems that an interaction of a plant's age and its competi-tive ability are paramount.

Regionalization

Much of the discussion concerning cyclic vegetative activity has been concerned with limited areas whereas the formations of Clements' monoclimax cover much larger areas of the world. Plant geographers have concerned themselves with very different scales of study and have used many characteristics of plants and vegetation for sampling. Phyto-geography includes both floristic and vegetational studies of areal pat-tern. Floristic characterization of regions has been attempted for a considerable time, Linneaus' Flora Lapponica being a classic example. At the present time geographers and taxonomists continue to determine the natural occurrences of taxa, and floristic systems are used to de-termine some of the larger units of plant life. Most of the present day floristic systems are somewhat similar, having 4 kingdoms: Holarctic, Palaeotropical, Neotropical, and Southern Oceanic, each with subdivi-sions (Engler and Diels 1936). The system in most widespread use in the English language is that of Good (1974) in which there are 37 prov-inces that are set up along lines of floristic coincidence and over which certain historical phenomena have left their mark. These are very much subjective classifications but Raup (1947) has attempted some degree of objectivity. The major use of these floristic provinces is that they can be seen as the raw materials out of which plant communities have been and are being assembled by short term changes in climatic and edaphic conditions; any boundaries drawn around the floristic provinces *per se* tend to coincide with abrupt ecological barriers.

Braun–Blanquet (1965) devised regional territories of six ranks based on a mixture of floristic and vegetational characters. His basic unit of study was conceptualized as the plant 'association' and associations are classified into groups of higher rank on the basis of relationships shown by species of high 'fidelity'. The 'Region' itself is Braun-Blanquet's most comprehensive unit containing many well defined climax communities with their own endemic families. Its unity is shown by species of high sociological importance which occur virtually throughout. Within Europe for instance there are parts of the Euro-Siberic–North American and Mediterranean regions. These are divided into 'Provinces' in which there is at least one climax community and various edaphic communities with their own endemic species and genera. The East and West Mediterranean provinces of Southern Europe are examples (Fig. 1). Below this are the 'sector', such as the British sector of the Atlantic province, with endemic species and special edaphic communities, the 'subsectors' with microendemics and geographical 'races' of the more widespread communities, the 'district' with its own characteristic communities and species and finally the 'subdistrict' in which a particular community of species is present or absent. This scheme of Braun–Blanquet is widely used in Europe and takes into

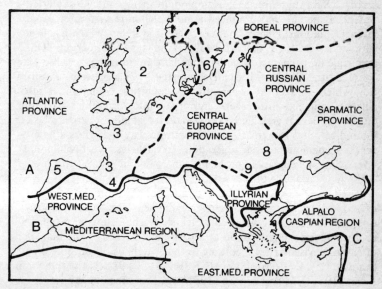

Fig. 1 Floristic-sociological territories of Europe (after Braun-Blanquet 1965)

account topographic, climatic and edaphic conditions both of the present day and of the recent past.

The Plant Community

Even the plant community, the term used as an expression of vegetative characteristics, has had a chequered history in the literature. It is best thought of as a generic term of convenience describing a sociological unit of any rank, occupying a real territory and having a characteristic composition and structure. 'Community-types' are the abstract or synthetic equivalent. The term community covers a wide spectrum of scale from the formation, such as the boreal forest or hot desert, the largest of all vegetation units, downwards. Plant groups are excluded from the concept if they lack integration of their parts. Some form of delimiting communities and grouping according to similarity is a necessity in order to produce a system for handling large quantities of complex data and to aid in understanding interrelationships.

The Scale Problem

Nearly all the early attempts to classify vegetation on a world scale were closely associated with features of the environment. Warming (1909) erected four major categories of vegetation: Hydrophytenveriene, Xerophytenveriene, Halophytenveriene, and Mesophytenveriene. Drude (1913) produced a more elaborate system of twenty seven formations. One of the more recent systems, which has been adopted by the IBP (Peterken 1967), is the idea of formations being mappable units characterized by dominant physiognomic, or life-form, combinations (Fosberg 1961). Dansereau (1957) suggested a classification which grouped biotopes into communities, communities into alliances, alliances into orders, and orders into formations. The four biochores: desert, grassland, savanna and forest are too large to be useful and respond to climate in a generalized way that gives a heterogeneous character to the vegetation. Nevertheless worthwhile and interesting biogeographic problems can be found even at this level (Hills and Randall 1968).

Very few maps of the land vegetation of the world use floristic data. Most use 'formation-classes' in the sense of Rübel (1930): Geographic units of vegetation which show characteristic responses to a particular climatic trend at a particular intensity. The number of classes devised varies according to the criteria of the authors. For example Hayek (1926) shows 16, Rubel (1930) 9, Schimper and von Faber (1935) 15, and Linton (1951) 39. Dansereau (1957) takes Schimper and von Faber's 15 as the basis for his chapter on Bioclimatology and compares them by climate, epharmonic responses, soil processes, ecological controls, and life-forms.

Perhaps the most important and most wide-ranging cause for correlation between species is their environmental requirement. Over very large areas where there are marked differences in species composition it is almost always environmental factors that are seen to correlate with community differences. These factors may be climatic, edaphic or hydrological. Within smaller areas, examined at a larger scale, these community differences are rarely as obvious and usually need quantitative rather than qualitative detection.

An hydrological example of the importance of only small variations in the environment is Harper and Sagar's (1953) work on the distribution of three species of buttercup on ridge and furrow grassland. They found that in permanent pasture *Ranunculus repens*, *R. bulbosus* and *R. acris* show a marked correlation with drainage, *R. repens* being most abundant in the old furrows, *R. bulbosus* on the crests of ridges, and *R. acris* on the slopes of the ridges. (Experimentally Harper and Sagar were able to conclude that the hydrological differences of ridge and furrow were critical at germination.) This, of course, was a relatively variable habitat to examine, but the results suggest that in apparently regular environments, slight variations in hydrology may be enough to influence species distribution.

This type of relationship to similar or dissimilar environments can be seen at very many different scales. In a study of the vegetation of the Monach Isles N.N.R. Outer Hebrides (Randall 1972) species of clover were seen to respond to variations in hydrology. On the machair grassland *Trifolium repens* was present in 67 per cent of quadrats examined and *T. pratense* in 31 per cent. Locations in which the latter species was present were always marginally lower. Examination of air photographs showed these lower areas to be old plough lines, though on the ground they were not deep enough to be followed. A more obvious relationship was between dune hollows and *Potentilla anserina* and *Odontites verna*, even though in summer there was no detectable difference in moisture content of the soil. The extreme in this series is between larger dune slacks and *Juncus articulatus* and *Agrostis stolonifera*. Soils in the slacks are permanently moister than the surrounding areas and actually waterlogged in winter. The dune slack vegetation would be obvious to anyone walking the area; the dune hollows would be obvious when the plants in question were in flower, and would show up if a large quadrat were used for analysis: the variations in the machair grassland would be missed unless a small quadrat were used.

On a very different scale the McGill Savanna Research programme

Vegetation Analysis

(Hills and Randall 1968) was able to show that in places along the savanna/forest boundary the continuation of grassland resulted from changes in soil structure and soil organic matter that resulted from burning and prolonged exposure to the extremes of climatic environment. Thus micro- and macro-environmental factors can be seen to affect both small patterns within the vegetation and community distribution. In all the cases described other factors such as competition may also be playing a part in vegetation distribution.

Competition

Competition is, in fact, the other major factor in pattern and correlation of species. This may be intraspecific, such as the well known decline in number of mature plants of a species relative to the number of germinating seeds per unit area, or interspecific. Studies of the relationships between woodland trees and their ground flora are good examples of the latter. Watt and Fraser (1933) were able to show that *Pinus sylvestris* played a major part in the development and growth of *Oxalis acetosella* and *Deschampsia flexuosa* below it. Conversely Hadfield (1957) has shown that young *Picea sitchensis* are adversely affected by the presence of *Calluna vulgaris* but beneficially affected by *Larix kaempferi*. The latter shades out the *Calluna*, roots at a different layer than the *P. sitchensis* and provides plant food in the form of an annual litter layer which the spruce makes use of.

According to Gauses' principle, competitive exclusion means that different species having identical ecological niches cannot exist for long in the same habitat. For this reason one rarely has many co-dominant species in any one community. Thus the general salt marsh community common to the north Norfolk coast is unusual in salt marsh succession. However, it has been shown that direct competition does not occur because phenology (time of flowering, fruiting, etc.) and rooting depth are different in the constituent species. One of the major problems of field studies of competition is that the correlation between competition and success or failure of a species can only be inferred. In most cases alternative assumptions can be made. It is only the ecological knowledge of the researcher that suggests cause and effect. All the data usually show is co-relation.

Environmental Change

Another important cause of correlation and hence the development of a community is biotic environmental modification. One form of this is succession, which was described above, but modification need not necessarily be successive. Along the East Anglian coast *Sueda fruticosa* may frequently accumulate low mounds of sand and organic matter which are colonized by *Honkenya peploides*, *Beta maritima*, *Cakile maritima* or other strand adventives. This community will only last so long as the *Suaeda* bush lives. Similarly one often finds *Crambe maritima* growing with *Glaucium flavum* on shingle beaches. Both grow in drift and their seeds may be buoyed to the location, but in other instances the curved seed pods of *Glaucium* have been seen to trap *Crambe* seed heads as they blow along the beach: young *Crambe* seedlings within mature *Glaucium* stands result (Scott and Randall 1976).

A different form of environmental modification by biota is the production of toxic substances by one species which inhibits the growth of all or most other species. This is most difficult to analyse because the effect observed might equally well be ascribed to competition. However, inhibitor substances do seem to occur and may well be more widespread than has previously been supposed. An early discovery was the production of juglone by walnut trees (*Juglens* spp.), which produces wilting on adjacent plant species. Similarly Randall (1970) found an impoverished community of *Fimbristylis cymosa* and *Caesalpinia bonduc* under *Hippomane mancinella* on the coasts of Barbados. The latter produces a toxic secretion which, when washed from the leaves, will scar human tissue.

Within this category of environmental modification by biota one should include the effects of animals and birds. Zooplethismic communities are frequently found on sea cliffs or coastal rocks where the excess quantities of nitrogen and other chemicals excreted in the guano favour such species as *Tripleurospermum maritimum* or *Stellaria media* but are toxic to most others. In woodlands starling roosting flocks may be so immense that trees will be killed and nothing but nitrophiles will grow below them until the birds move to a new roost. Such effect may be local or widespread and may recur frequently enough to cause distinct and recognizable communities.

Kershaw (1973) makes the point that morphological scales of pattern will always be present simply because every plant species has its recognizable pattern of growth. Also environment is always variable enough to produce one or more scales of environmental pattern. Competition and modification of one form or another are also widespread so that pattern will always be present in vegetation at a variety of scales, although the causal, rather than correlative, factors will often be difficult to elucidate. At the morphological and small pattern scales methods of detection are simple since one is dealing with the individual specimens of a species. However, at the other scales there are many problems in the detection of natural communities—what does one sample and how does one analyse the results? Sometimes these methodological problems have seemed so great that researchers in vegetation studies have been more concerned with technique than results.

33

5 Techniques in Sampling Vegetation Data

The requirements

The vegetation mantle of the earth is extremely complex and virtually continuous over much of the landscape. Thus all vegetation data gathering must be selective since it would be impractical to undertake any total recording or description of vegetation. This results in two considerations being of prime importance: the type of data to collect and the spatial location of the sampling points. Before any method of vegetation sampling is used in the field a large number of considerations must be examined. Firstly, the primary aim in making the data collection must be appreciated since too little or too much data and a lack or excess of detail must be avoided. A decision must be made concerning the subjectivity or objectivity required. Frequently this resolves itself according to the qualitative or quantitative nature of the synthesis. Secondly, considerations both of time and expense are needed. Both are usually limited and the normal requirement is the most useful information with the least time and money spent. Thirdly, the nature of the vegetation cover must be assessed. If it is particularly homogeneous, then the questions required to be answered must *de facto* be quantitative in nature since the observable differences will be slight. Conversely if the vegetation is of a heterogeneous nature, then there may be obvious qualitative differences, although these need not be the most important or even the most interesting ones.

Physiognomic Analysis

Within biogeographic literature there has been considerable discussion regarding 'what should we map' (Fosberg 1961). Physiognomic (superficial appearance) analysis has been put forward strongly by certain biogeographers because of the human use of such information and because of the ease with which an untrained layman can both observe and record (Dansereau 1957, Dansereau, Buell and Dagon 1966, Küchler 1949). Structure is a visibly outstanding feature of the vegetation and may rank before composition in a description of landscape. Structure does not always vary with composition but it does generally change with climate, topography and soil, and with successional stage. Kellman (1975) considers that structural descriptions are of limited use except in such roles as military operations, but in reconnaisance or the

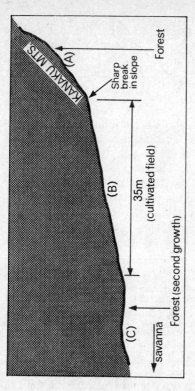

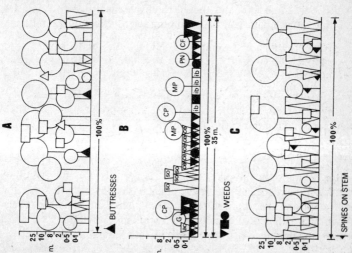

Fig. 2 The relationship of the upper fields of cultivation at Kumu Guyana, to the topography and forest-savanna boundary 1967.

An abstract structural description of one of these cultivated fields is shown in ('B') while similar descriptions of the immediately adjacent forest is shown in ('A') and ('C'). The method of description has been slightly modified from that of Dansereau, Buell, and Dagon (1966) in order to accommodate the cultivated plants: pawpaw (CP), plantains (MP), sugar cane (so), cotton (G), black pepper (PN), hot or chili pepper (CF), pine-apple (ce), eddoe (ac), and sweet potato (ib).   LEGEND: ○ Erect woody plants; □ Decumbent or climbing woody plants; △ Epiphytes; △ Herbs.

primary survey of considerable areas in a short period of time, they have been found invaluable, especially as part of a geographic synthesis. Fig. 2 illustrates the way in which a structural diagram may be used to relate cultivation to topography and adjacent natural vegetation.

Leaf Characteristics

A more sophisticated form of structural description is Raunkiaer's (1934) use of life-form and leaf-size. These systems were pre-dated by many authors right back to Humboldt (1805) but it is only the Raunkiaer systems that have received world-wide use because of their simplicity and homogeneity which has allowed statistical treatment.

Raunkiaer developed five principal life-form classes, arranged according to increased protection of the perennating (overwintering) buds: phanerophytes (trees and tall shrubs), chamaephytes (trailing shrubs), hemicryptophytes (herbs), geophytes (bulbs) and therophytes (annuals). The percentage of each life-form class in a flora can be compared with Raunkiaer's normal spectrum, and deviations measured in relative terms. Cain (1950) has examined in depth the spatial distribution of life-forms. Chamaephytes are seen to increase markedly with altitude, whereas theorphytes decrease. The fifteen formations of Schimper and von Faber (1935) can be characterized by their life-form classes and a general consistency is revealed between climate and life-forms in the great world systems of vegetation, Fig. 3.

The leaf-size classes of Raunkiaer are based upon the observation that leaves are large in moist tropical vegetation and decrease in size as conditions become drier and cooler. He divided leaf size into six classes each nine times larger than the minimum twenty five square millimeter.

1	Leptophyll	up to 25 sq. mm	4	Mesophyll	up to 182.25 sq. cm
2	Nanophyll	up to 2.25 sq. cm	5	Macrophyll	up to 1640.25 sq. cm
3	Microphyll	up to 10.25 sq. cm	6	Megaphyll	larger than 1640.25 sq

These classes may be very tedious to use though most leaf areas approximate to two-thirds the area of a rectangle proportionate to their length and maximum width.

Species Characteristics

More commonly the plant species has been used as the sampling unit. Most plant species are clearly distinguishable in the field (with a few exceptions like *Euphrasia* spp. or *Epilobium* spp.) and are consistent in their behaviour towards the environment reflecting its changes in a varying manner. Also they result in data that are easily handled in a statistical fashion (Perring and Walters 1962). The most objective

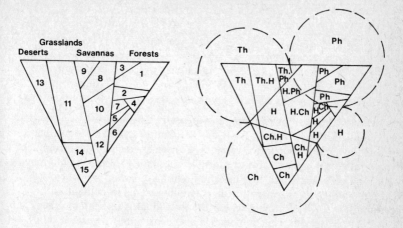

Fig. 3 The formation classes of the world and their relations to Raunkiaer's (1934) life-forms: 1. Tropical rainforest; 2. Subtropical rainforest; 3. Monsoon forest; 4. Temperate rainforest; 5. Summergreen deciduous forest; 6. Needle-leaf forest; 7. Evergreen hardwood forest; 8. Savanna woodland; 9. Thorn forest; 10. Savanna; 11. Steppe and half-desert; 12. Heath; 13. Dry desert; 14. Tundra and cold woodland; 15. Cold Desert.

measurement of a plant species is whether it is present or absent, but this allows for no assessment of the importance of that fact, nor does it allow for examination of the function of the vegetation community. Frequently this will be the prime reason for vegetation study. Thus the most useful data to collect are some measures of the importance of each species. What measures are used will normally depend upon resources available, specific objectives and species morphology. Ideally counts based upon individual specimens are the most informative data but with grasses, bracken or similar species this would be impractical. In forests studies may be required that emphasize timber volume or tree regeneration. These would obviously require different data.

Types of data—estimations

The most basic data one can collect are the numbers of individuals of each species in an area. If these numbers are estimated they are generally called abundance and classed in a few abundance classes, five being commonly used:

Vegetation Analysis

Abundance Class	Tansley and Chipp (1926)	Hanson and Love (1930)	Braun–Blanquet (1965)
1	Rare	Very scarce	Very sparse
2	Occasional	Scarce	Occasional
3	Frequent	Infrequent	Not numerous
4	Abundant	Frequent	Numerous
5	Very abundant	Abundant	Very numerous

Obviously estimations of these sorts are extremely subjective and errors creep in because of colour, form, aesthetic interest, ease of identification etc., as well as operator variance. Conversely density data (numbers per unit area) is almost always too time consuming to collect, and may be rendered inaccurate by suckering (shoots coming from below ground), layering (propagation from trailing stems in contact with ground), or other forms in which the individual is inseparable from a visual inspection. Thus abundance and density must be used with caution. This is not to say that they are useless. With certain life-forms they may be effective. Acocks' (1953) study of veld types is apparently an example of successful use of this type of information. One of the insidious problems with abundance/density as with many other forms of vegetation data collection is variation with time. Tree seedling data from any one year may be highly misleading since most species have good and bad 'mast' years. Likewise visual evidence of a herb's existence may vary from season to season or month to month, as with the spring annuals of sand dunes or the desert therophytes.

Scales

Numbers data have sometimes been integrated into scales including other quantitative measures of community structure, such as the cover-abundance scale of Domin (Table 2). Like abundance classes these scales are notoriously difficult to use accurately (Hope–Simpson 1940), especially the Domin with its unequal class values. As Kellman (1975) has said numbers assigned to ranks cannot legitimately be used in statistical computations but Bannister (1966) has shown that, with care, subjective estimates may be used as the basis for classification and ordination. Particularly where extensive surveys are carried out in a relatively short time, abundance and/or cover classes are often the best method of data gathering despite their shortcomings. Such surveys will frequently be the work of the biogeographer.

TABLE 2

The Domin Scale of cover–abundance

+	Occurring as a single individual with reduced vigor; no measurable cover
1	Occurring as one or two individuals with normal vigor; no measurable cover
2	Occurring as several individuals; no measurable cover
3	Occurring as numerous individuals but with cover less than 4% of total area
4	Cover up to 1/10 (4 to 10%) of total area
5	Cover about 1/5 (11 to 25%) of total area
6	Cover 1/4 to 1/3 (26 to 33%) of total area
7	Cover 1/3 to 1/2 (34 to 50%) of total area
8	Cover 1/2 to 3/4 (51 to 75%) of total area
9	Cover 3/4 to 9/10 (76 to 90%) of total area
10	Cover 9/10 to complete (91 to 100%)

Dominance

Cover alone is a feature of the vegetation associated with dominance. It is an estimation of the area of foliage of a species projected vertically down to the ground. Again it is a crude and subjective measure but is usually put into five or less classes which give considerably more information than presence alone. A particular use of cover estimates is by stratum in a multi–layer community: this will give considerable insight into the vegetation, because species transgression will show up.

Some confusion arises concerning the dominant species physiognomically and the dominant species within the total community. Often these are one and the same such as the oak trees in an oak forest. In the savannas of South America the sandpaper trees are the physical dominants but the grass or sedge–dominated field layer exercises the primary control in the ecosystem. Uusually the number of dominant species in a community varies with environmental quality. Over much of the temperate world there is only one dominant but in tropical areas the better abiotic conditions the more co-dominant species there are.

Foresters use dominance and cover in a slightly different manner. Dominance is always related to presence in the superior layer of the forest canopy and cover is estimated from basal area at breast height. The latter is a measure of stump area if the tree were cut down at 4.5ft (1.5m). Range managers use the term 'basal area' for the area of grasses at utilization height. This may approximate cover with large grass clumps such as *Panicum turgidum* in the Negev but the difference may be enormous where several small clumps occur together.

Vegetation Analysis

Cover

Cover may be estimated quantitatively rather than visually by means of point quadrats (Goodall 1952), frames of pins which are lowered one at a time and the species touched by each pin recorded. The accuracy of this method though is dependent upon pin diameter. For comparative purposes this problem is irrelevant but in absolute studies optical cross-wires are the ideal. Obviously such methods are very tedious and time-consuming and are only worthwhile for a really detailed survey of a small area, and then only for low-growing vegetation.

If the prime interest in cover is a study of vegetation productivity, and damage to the plot is immaterial then a measure of the actual weight of plant material—the standing crop or biomass (dry weight per unit area) may be best. This is usually limited to above-ground parts of plants and is expressed as dry-weight of tissue per unit area.

Frequency

The measure of species that is perhaps most fraught with difficulty is frequency: the proportion of sample units which contain a given species. The measure is simply obtained by recording presence or absence within a series of quadrats, such that a species would have a 25 per cent frequency if it occurred in 50 quadrats out of 200 thrown. Local frequency may be obtained by sub-dividing each quadrat and averaging the frequency within each quadrat. The usual forms of frequency used are shoot frequency or root frequency. The latter requires the plant to be actually rooted within the quadrat. Thus errors in estimation compared with cover or abundance are small but the frequency figure may be seriously distorted by quadrat size, plant size and spacing of individuals. Unless this is appreciated by the researcher, the frequency class to which he attributes his species may well be an entirely artificial product. On the other hand frequency may be used to measure size and spacing so long as quadrat size is kept constant.

Quadrat size must always be stated when using percentage frequency. In this way users of the data may appreciate its limitations. Plant size is only important if one is using shoot frequency since large individuals will inevitably be over-represented. Thus, so long as the frequency measure used is stated the data can be used by other workers. Frequency measures the degree of dispersion of a species throughout the area sampled and thus complements density. Two species may have similar densities in an area but one be dispersed and the other contagious in distribution. The former will have a much higher frequency. The relationship between frequency and density is not proportional because random dispersal so rarely occurs in living organisms.

Raunkiaer (1934) developed a law of frequency based on five 20 per cent frequency classes. This states that the numbers of species of a community are distributed in the classes in the following manner: A (1-20) > B (21-40) > C (41-60) ≥ D (61-80) < E (81-100). The increase in size of the 81-100 per cent class has been the basis of definition of many plant associations but it is in fact caused by a statistical property reflecting the theoretical infinite range of density. Raunkiaer's law of frequency has been criticized by many researchers on the basis of quadrat size and sample number.

Synthetic measures

Obviously there are many other ways of describing species representation in a community but most of these will be concerned with a particular topic such as how well a species is growing in a particular environment. There is often good reason for choosing one's own descriptive technique to suit the purpose of the research so long as it is realized that the more commonly used measures are more suitable for comparison and that, by and large, their pitfalls have been discovered. One answer that has been suggested to the problem of description is to develop a synthetic index that incorporates many of the other measures. The best of these is Curtis and McIntosh's (1951) 'importance index', combining frequency, density and basal area on a scale of 0-300. This has its problems in that it assigns arbitrary ranks to each measure but in empirical terms it has the advantages of rapidity and comprehensiveness: no measure is perfect.

Minimal Area and Quadrats

When the researcher has resolved the problem of what to sample he is still faced with the questions of how and where to sample. These questions may involve convenience or accuracy of representation, or, as we saw with frequency, may affect the measure used. One of the primary factors of interest is the species-area relationship, the fact that the number of species present increases as area increases within any given community. This relationship is both a community characteristic, involving the law of diminishing returns, worthy of measure and also a useful determinant in developing a sampling programme. The relationship between minimal area and homogeneity will be dealt with later. Species-area curves can be produced by sampling a small area then progressively doubling in size and noting the additional species. Eventually the production of nested plots will add only a small number of new species. The resultant curve from plotting mean number of species per plot against size of sample plots will enable 'minimal area' to be suggested: that is the smallest area that provides suitable conditions for a particular community to develop its essential combination of species. Minimal area can never be assessed accurately because the species–area curve has no distinct inflexion for most communities. However, minimum quadrat area and minimum quadrat number for sampling any given community can be usefully inferred. Most work has tended to show 1–2 sq. m to be suitable for herbaceous vegetation, about 4 sq. m for low shrubs, 16 sq. m for small trees and 100 sq. m for forest.

By tradition the shape of sampling quadrats has been square but there are practical advantages in the use of rectangular or circular forms. The most severe problem encountered is the edge effect of very small or long, thin quadrats. These, therefore, should be avoided.

Sample units need not be quadrats. One alternative is the line intercept, a record of the species that make contact with a line of, say, string or wire stretched through a community. This sample may have a dimension added by making it a bisect or profile in which vertical relations of the vegetation (above and/or below ground) may be recorded. Another variation is the point sample, a unit which removes the whole problem of area. Foresters have sometimes used plotless sampling to obtain basal area information (Cottam and Curtis 1956) but their techniques have not become widespread. Because of their concern with space, geographers may well see increasing use for methods such as nearest neighbour (Cottam *et al.* 1953).

Sampling Designs

Whether one is sampling by means of a series of individual quadrats, lines of quadrats (transects), line intercepts, points, or other variants, one must consider the arrangement of the sample units within the community to be studied. Most quantitative ecologists recommend random sampling because the results will be objective and therefore amenable to statistical manipulation. Random is a mathematical concept that involves the use of random numbers (or cards, dices, etc.) to locate samples. It does not mean location by closing one's eyes or throwing the quadrat frame over the shoulder. These methods are not random and the results have been proved to be biased. However, from a spatial point of view random sampling may not always be satisfactory. Unless the sample is extremely large there is a strong likelihood with random sampling that certain areas will be under-represented. Botanically this may be unimportant but for mapping or other spatial analysis good aereal coverage is vital. Therefore other methods may be used. The most common are the regularly spaced or grid samples which are deliberately located to ensure total coverage of an area. So long as the co-ordinates of the grid are chosen randomly this becomes in effect a random sample. Kershaw (1973) raises the problem of the chosen grid coinciding with ridge and furrow on old ploughland and Shimwell (1971) discusses the possibility of harmonizing with periglacial stripe patterns. These are theoretic rather than common practical problems. If met, they can be deliberately overcome by changing the scale of the sampling net, using the grid intersections as origins for nested samples, or choosing random samples from within the grid squares. Similarly if one's grid scale is seen to be too coarse to pick up some micro-communities nested samples can be used.

Transects

A special example of regularly spaced samples is the transect. This is a line of contiguous or regularly spaced samples deliberately placed along a known topographic or environmental gradient to sample change. So long as the location, as distinct from the direction, of the transect is randomly chosen then the samples obtained can be statistically treated. This method was employed to sample the beach vegetation of Barbados (Randall 1970) where the effects of the sea were considered to be the prime environmental parameter.

Sampling is not only concerned with space but also time. Sample plots may be temporary or permanent. Most sample plots are temporary but it is often important to observe and record vegetation changes over time. Permanent sample plots must have their location marked on the ground or tied into some permanent marker that can be relocated with accuracy. 'Permanent' may mean a season, a year or even a decade or more. Permanent quadrats may be of any size and may be constructed as 'exclosure quadrats' if grazing interference is to be avoided.

Vegetation sampling does not *have* to be objective. Far too much of the literature ignores this fact. Only objectively collected data can by statistically analysed but for descriptive purposes the small sample, subjectively chosen to exemplify the vegetation of an area, is far better than a small random sample which neither fully describes nor has enough data to manipulate.

The most important consideration in constructing a sampling design is the problem under study. Within limits of scientific accuracy it will be this, rather than theory, which decides the logistics of the survey.

6 Techniques of Analysis

Whatever the form of data collection used, the resultant information is usually in the form of a two-way table listing site and vegetation data with presence–absence, semi-quantitative or fully quantitative data within. The objective of analysis is to simplify the real-world situation, discarding the 'noise' or useless information and keeping the hard core. With the more simplistic model a greater understanding of vegetation form and function should be possible so that one can predict vegetation type in both time and space dimensions. Such an objective presupposes non-randomness and some tendency towards spatial order within vegetation. The dichotomy of views within the organismic/continuum debate outlined earlier (Whittaker 1953, 1967) and the detection of non-randomness (Greig-Smith 1964) have resulted in a disparate attitude towards the amount of spatial order that exists. On the one hand this has resulted in classification: the *segregation* of individuals into groups along with others which are most similar and apart from those which are less similar: on the other ordination: the *arrangement* of units in uni- or multi-dimensional space. As will be seen the two approaches need not be mutually exclusive. Both aim to extract the maximum information concerning species relationships in the minimum time.

Phytosociology

One of the most widespread techniques is the European floristic system of vegetation analysis. There are three schools, the Raunkiaerian, the Uppsala and the Zürich-Montpelier. Shimwell (1971) devotes a useful chapter to these schools. Only the most important, the Zürich-Montpelier (Z-M) will be covered here. This system was originated by Braun-Blanquet (1965). Work carried out using this widely accepted classificatory system is aired at the annual symposia of the International Society for Plant Geography. Much of the work has been described in French and German and English-speaking contributors are few. Thus confusion over the methodology has been widespread in Britain and North America. In fact Kellman (1975) puts the extreme view:

In essence, the system appears to possess most of the undesirable attributes of an organization and few of the desirable features of a summary. It is highly selective of the data it treats, employs ill-specified strategies and has proven impossible to apply in areas of floristically complex vegetation. Above all it relegates the user to a role which is little more than that of a descriptive technician. Its continued use in vegetation studies appears to reflect more the inertia of the system than its intrinsic value.

Becking (1957), Küchler (1967), Moore (1962), Mueller-Dombois and Ellenberg (1974), Poore (1955) and Shimwell (1971) have helped to widen English knowledge of the system and in the last two decades its use has spread somewhat into English speaking areas. Users of the system initiate several stages of work. First uniform areas are chosen for description. This in itself precludes any objective statistical treatment thereafter, but is not *de facto* an error. Certainly in primary sampling of an extensive area a more representative picture may be gained thereby than by making use of an unsatisfactorily small random sample.

The Z-M Scheme

Within each uniform area randomly chosen several relevés are made. These are complete plant lists within a given sized quadrat, each plant being given subjective weightings and sociability classes. Habitat features, vegetation height and other relevant data are collected (Table 4). On grassland a relevé takes about twenty minutes, in woodland forty five minutes. When the relevés are completed the synthetic stage begins. This is the delimitation or classification of the vegetation units by visual comparison of relevés entered into a table i.e. pattern picking. This exercise is carried out by first making a raw data matrix which is then progressively rearranged by detection of differential species. Species are considered to be constant (in 60 per cent or more relevés), rare (in 10 per cent or less relevés) or intermediate. Among the 'intermediate' there will be correlated species which, when detected, will enable a rewrite of a 'partial table' with the relevés more ordered. This will be done several times until a final editing is done. Partial tables are a classification process. The final table is an ordination bringing the most similar relevés side by side along an ecological gradient. This results in vegetational units characterized in blocks of differential species. The criticism that this exercise is time consuming and purely technicians' work has now been overcome by the use of computers. Céska and Roemer (1971) and others have published suitable programmes.

The validity of the units obtained are tested by a return to the field and if some areas are unresolved then further relevés are taken and added into the table so that the *total* vegetation is mappable.

Vegetation Analysis

Hypotheses can be generated concerning correlation of vegetation units and ecological factors. These can be visually tested on the table by reference to the environmental data. Strictly statistical treatments may not be made although some workers use χ^2.

TABLE 3

A sample relevé

RELEVE No. 37   LOCALITY: Stockay, Monach Isles   DATE: 19.6.69

GRID REFERENCE: NR 662632   ALTITUDE: 4.5 ft

VEGETATION	Semi-stable dune herb-grassland, 0.6 m tall; ungrazed but evidence of eider duck and fulmar nesting.
COVER	Herb, 70%; Moss 20%; Litter 30%; Bare sand 30%.
SOIL TYPE	Calcareous sand with little incorporated humus.
SPECIES	7 *Heracleum sphondylium*
	5 *Armeria maritima*
	4 *Plantago lanceolata*
	4 *Silene maritima*
	3 *Festuca rubra*
	3 *Galium verum*
	2 *Leontodon autumnalis*
MOSSES	5 *Tortula ruraliformis*
	+ *Hypnum cupressiforme*

The vegetation units of the final tables can be fitted into the general hierarchy developed by the phytosociologists, based on Braun–Blanquet's associations, alliances, orders and classes. Like any classification its main criteria is its usefulness. Many of the criticisms should be levelled at bad practitioners rather than at the technique itself. Real criticisms of the technique result from the prejudice of quantitative ecology, in thinking that all study must be quantitative science rather than a landscape art.

Classification

The forms of classification are four-dimensional: hierarchical or reticulate, divisive or agglomerative, monothetic or polythetic, and qualitative or quantitative. Most ecological classifications are, like the Z-M, hierarchical but in recent years the English-speaking proponents of the subject have attempted to be more objective and truly quantitative. How far they have succeeded is open to question. Monothetic (single factor) divisive methods have been the most frequently used, originating in the work of Goodall (1953), and developing through the 'Southampton school' of association analysis (Williams and Lambert 1959, Lambert and Dale 1964) to a mass of alternative statistical techniques, using different mathematical models (Crawford, Wishart and Campbell 1970, Randall 1972). Divisive techniques, by their very nature, must be hierarchical, starting with a single heterogeneous population and dividing down to a number of virtually homogeneous groups. Monothetic systems of division are the simplest because separation of the two most similar groups at any stage is only in terms of *one attribute*.

Division

MacNaughton-Smith *et al.* (1964) and Edwards and Cavalli-Sforza (1965) have attempted to use polythetic (multiple factor) divisive methods. These are infinitely more complicated since groups are separated or held together by a coefficient of overall similarity or dissimilarity. MacNaughton-Smith's dissimilarity technique asked the question of each member of the population 'How abnormal are you?' As a result a split is caused between a home block and a splinter group. These are accepted as two populations and the process is repeated. However, Edwards and Cavalli-Sforza showed that with a large population the number of calculations required soon make the method impossible to use.

A geographical advantage of divisive hierarchies is that groups formed at any level within the hierarchy can be mapped and compared with environmental features. All techniques are similar in that they use specified mathematical models with properties fixed before classification begins. Thus, results are replicable and therefore objective. However, there are several decisions that have to be made. In systems including χ^2, do you operate with or without Yates' correction? Do you include all collected data on the initial 2×2 contingency table or do you save on computation by excluding the rarities? What do you do about equal divisions? Even the decision: 'Which coefficient do you use?' involves subjective decision. Slight differences such as chaining effects result from the use of different coefficients and it is a problem to know how to treat them. Perhaps one of the greatest problems is that atypical quadrats occur in nature and a monothetic system can so easily lead to misclassification thereby.

Williams *et al.* (1966) have reviewed the possible alternative tech-

47

niques on which classification may be based. They consider hierarchical methods, which optimize the route of analysis, preferable to reticulate methods, which optimize the end result. Historically this view is upheld but it has little logical basis. The rank is of considerable importance but it is the final groups that are of most interest to biogeographers and methods must be judged on the spatial meaning of these. Hierarchical methods do have the advantage that further information can be the more easily assigned to the appropriate class.

Fusion

Williams *et al.* (1966) favour agglomerative methods over divisive. Since monothetic agglomerative processes are impossible, this leaves polythetic fusion. In this process all individuals are first considered separately and the msot similar pair are fused into a cluster; the process is again repeated to build up the hierarchy until only two clusters remain. Virtually all monothetic division generates some unusual groupings due to the constraint of presence/absence used to determine cluster membership. Many forms of polythetic fusion need not suffer from this constraint and quantitative or semi-quantitative data (such as Domin values) may be used.

A popular statistic for use in polythetic fusion is 'increase in error sum of squares' (ESS) as proposed by Ward (1963) and used by Orloci (1967). ESS is defined as the sum of the distances from each individual to the centroid of its parent cluster. Ward proposed the hierarchical method which combines those two clusters whose fusion yields the least increase in the error sum. Fig. 4 shows an example of a classification dendogram derived with this method and also the geographical distribution of the eight categories used for descriptive analysis. It will be noticed that for most steps of the fusion the increase in error sums of squares is extremely small. At later stages the increase becomes larger, representing the uniting of unlike groups. It is just before this stage that biogeographically relevant clusters are most frequently produced.

Misclassification

A classification strategy such as that discussed results in groups of quadrats for which the membership of the group is defined by a vector of conditions. Usually the characteristics of a group are determined by the presence of a large number of species, and generalization leads, almost invariably, to a misclassification of some quadrats. This situation will occur when the overall characteristics of the group are met but one particular condition is not. Several procedures have been devised to overcome such misclassification (e.g. Lance and Williams 1967).

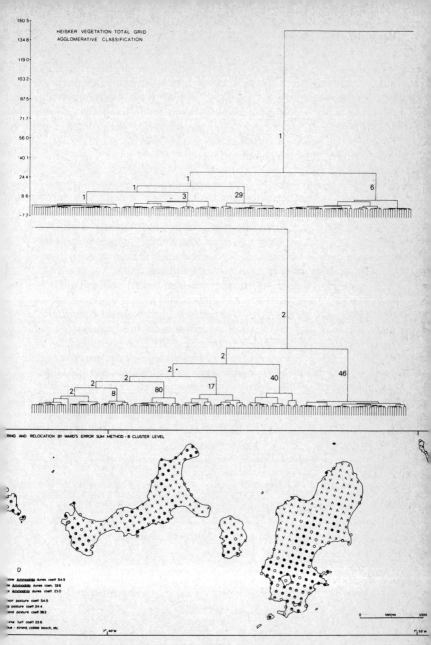

Fig. 4 Classification dendogram of three hundred and thirty-six stands of Monach Isles vegetation produced by polythetic fusion using Ward's (1963) 'error sum' statistic, with map of the 8 cluster classification.

49

Vegetation Analysis

In such exercises limits have to be set both to the minimum size of parent groups so that trivial classes are not produced and also to the number of computational iterations, in order to avoid oscillatory conditions in which a quadrat changes from one group to another consecutively. Biogeogrpahically, relocation of this type means that certain borderline or 'ecotonal' quadrats will change allegiance at the different significant levels of the computation.

Objectivity

Because different classification structures can be derived from the same data using differing models, e.g. Fig. 5, it is obvious that their objectiveness is called into question. This is a result of site selection and coefficient being subjective even though objectivity is present in the way the model is applied. Thus, even when the categories derived correlate closely with environmental gradients, one is not proving a relationship but predicting one that has then to be tested in field or laboratory experiment. Thus, as Lambert and Dale (1964) suggested, the real role of classificatory systems lies in hypothesis generation, and not hypothesis testing, as one might suspect from much work published over the last decade. Yarranton (1967) even goes so far as to suggest that it is only by applying the 'untenable' organismal concept of vegetation that one can legitimately formulate and test vegetation hypotheses, rather than analyse species by species. Since the truth lies somewhere between the extremes of the individualistic and the organismal concepts, so a limited amount of analysis and prediction can result from classification. Equally, or even more important, is the role

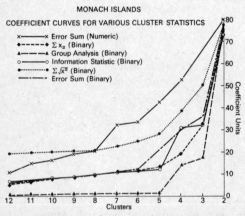

MONACH ISLANDS

COEFFICIENT CURVES FOR VARIOUS CLUSTER STATISTICS

x———x Error Sum (Numeric)
◆-----◆ Σx_2 (Binary)
▲———▲ Group Analysis (Binary)
○———○ Information Statistic (Binary)
●·······● $\Sigma \sqrt{x^2}$ (Binary)
—·—·— Error Sum (Binary)

Fig. 5 Coefficient curves for various cluster statistics tested with vegetation data from the Monach Isles. All coefficients have been converted to a basic scale to make the curves directly comparable.

of classification in the synthesis of information known about a particular piece of vegetation.

Ordination or Classification

An alternative method of treating vegetation data is ordination. This method has often been treated as being directly in opposition to classification since it is a multivariate technique concerned with between-group similarities. The individuals (quadrats) are arranged on axes with their properties (species) determining their positions. The structuring, or simplification, is not in this case necessarily organismic or spatial. An early exercise in this technique is illustrated by Beals (1960) who was associated with the Wisconsin School. Another example of relevance to the biogeographer is the work of Colebrook (1964) on the geographical distribution of zooplankton. It is now a commonly used method especially by researchers who have studied at Bangor (Orloci, 1966, 1967 *inter alia*). Ordination, like classification, stresses orderly tendencies in data but instead of segregating data into clusters it produces clines. If the data were a completely random assemblage of species then an isodiametric array of points would result. Thus, as Lambert and Dale (1964) assert there is no basic reason for ordination to be associated with continuous data any more than classification with discontinuous data.

Much of this fundamental controversy is a result of regarding vegetation separately from the ecosystem within which it occurs. The holistic character (Stoddart 1965) and the effects of alterations in the ecosystem as a unit have not been fully appreciated. Anderson (1965) points out that environmental selection acts with differing intensity upon the vegetation cover but the resultant 'communities', whether diffuse or discrete, are functional entities and their actual boundaries are relatively unimportant features. A similar conceptual idea can be seen in the human geographic development of the region.

Webb (1954) foresaw the futility of this dichotomy between classification and ordination but little was done about it in practice until the work of Barkham (1968) and Norris (1968), who used both techniques and attempted to compare their efficacy. Grieg-Smith (1961) theorized that the ordination of classified data would be an efficient means of operation, and Gittins (1965) and Randall (1972) practised this to a limited extent.

Ordination

Early attempts at ordination were those of Curtis at Wisconsin (Brown and Curtis 1952). This involved the ordering of sample sites along a single axis, using as data the 'importance values' (a synthetic measure) of the most abundant tree species at each site. If the abundances of other species are plotted along the same axis, population changes will show up, often as quite regular curves. Environmental variables were frequently seen to describe similar curves to given species. Bray and Curtis (1957) developed the ordination technique further by using a weighted floristic similarity index calculated between all sites to derive a site similarity matrix. From a reciprocal of this matrix ordination axes were extracted geometrically, using the most dissimilar stands as end points of the axis. Using this technique three orthogonal axes can be extracted, giving a crude reduction of the total data array. Bray and Curtis found that the ranking of sites along their axes could be related to environmental factors, the major one in this case being successional recovery of the vegetation from past human disturbance. It was concluded by the Wisconsin school that the results of their analyses illustrated the continuous nature of vegetation change in space and the absence of discrete associations. What is not clear is that although continuity in composition can be demonstrated, such non-spatial data cannot demonstrate continuity in the real world. Orloci (1966) questioned the fundamental geometric accuracy of Bray and Curtis' work and later developments in ordination have drawn upon factor analysis and principal component analysis.

Factor analysis assumes that there are underlying factors to which most species will be responding and which may be identified if species gradients occur. Individualistic species variation is noise that is eliminated from the analysis. Rummell (1968) clearly describes this technique. It has been little used in vegetation analysis because of its conceptual limitations. Principal components analysis on the other hand partitions the variance within rows of a data matrix into new variables. There are as many of these as there are rows of data but the first principal component has the largest variance that can be found by a linear transformation of the data. The second has the largest variance orthogonal to the first, and so on. Thus, it is hoped that the bulk of the variance in the data will be explained by the first three or, at the most, five components.

Hence principal components analysis defines a space in which the relationships between individuals, or clusters produced by classification, are represented by distances. The two principal eigen vectors obtained show the best two-dimensional representation of these relationships. Further vectors may be drawn to describe less important relationships. A procedural summary is presented in Mather and Doornkamp (1970).

Austin and Orloci (1966), Orloci (1966), and many others have applied a principal components analysis to an M × N vegetation matrix to obtain an efficient representation of its structure. In most cases a relatively small matrix was used since the computer time involved was considerable. Orloci (1967) was using most of the available computer storage space with 39 samples and 109 species.

An effective answer to the problems of computer time and storage space is to reduce the dimensions of the matrix to be ordinated by first obtaining groupings in a classification technique such as error sum of squares. It is true that the generality of N-dimensional space is sacrificed for K-dimensional space (K being the reusltant number of groupings) but it is doubted whether the inaccuracies so derived are any greater or more significant than those discussed by Orloci (1966) and Bannister (1968). Naturally this method is not regarded as a projection of the N vectors, but as a subjective mapping of the space N, in K. Furthermore, the reason behind ordinating vegetation data is often to obtain a visual impression by graphical representation of the relationship between the quadrats of a survey. Although the use of Euclidean distance (Orloci 1966) and N-dimensional space, for example, are more statistically elegant, it is questionable whether they are any more meaningful than the 'C' coefficient of Bray and Curtis (1957) or K-dimensional space, in terms of the biogeographical information gain. Ordination is limited in its role as nothing more than a hypothesis generator, because there is a built-in assumption of linearity in species correlations with influencing factors, and of orthogonality of the axes extracted. However, in the absence of truly viable alternatives, classification and ordination provide some means of simplifying the real-world.

The interesting result is that in nearly every case such procedures give the information which any trained man will derive from his knowledge of factors external to the data, and that in most cases they will find something that the biogeographer, despite his extra knowledge, has missed. It seems that the real value of numerical analysis is that it does automatically much that would otherwise have to be done by the trained man, leaving him free to apply his special knowledge to extend the analysis.

Hypothesis Testing

Hypotheses formulated by classification, ordination, or, in fact, by any other method, must then be tested with further data not used in setting them up. Virtually all hypotheses in vegetation analysis seek to correlate trends in the vegetation with trends in one or more environmental factors. If these correlations are investigated indirectly via the vegetation patterns alone, then unconvincing logic may result. Whittaker (1967) attempted to clarify this type of hypothesis testing via 'gradient analysis'. This involves the subjective selection of composite, environmental gradients along which the response of the vegetation is examined and plotted in terms of structural features, species richness or even individual species abundance. Loucks (1962) attempted to use a similar methodology combining the three environmental variables of moisture régime, nutrient status and local climate to derive synthetic gradients by means of scalars.

In those areas where the trends in the real–world environment are relatively simple and where the research worker has considerable experience, an environmental approach to vegetation can often produce an applied or predictive result. For instance, in a coastal ecosystem where the influence of the sea is paramount in environmental terms, histograms of cover percentage of the common (or all) species in each quadrat along a transect orthogonal to the shore can be plotted, then superimposed. These histograms indicate that although no two species have identical distributions, 'community types' or zones having dominant species can be recognized. Using this method Randall (1970) demonstrated that similar groups of sequential zones are present in each of the three coastal ecosystems of Barbados. This information was then used in order to predict the most suitable species for landscape planting on a new coastal road. In this particular example the vegetation associations were not only seen to be associated with distinct geomorphic units but were also significantly correlated with aerial and ground salt quantities, and $CaCO_3$ and organic matter content of the substrate.

Obviously such techniques are not total explanations of vegetation distributions but are synthetic 'first approximations' (Poore 1955) which can then be used for further analysis. This is a standard scientific method of advance. In spatial terms such methodologies are limited to relatively small areas of straightforward or well-understood ecosystems. Neither equatorial forests nor trans-continental trends would respond more than crudely to environmental analysis. However, at the small spatial scales the techniques of ordination and classification already described can be applied. One example would be to plot up vegetation classification clusters on other diagrams which are drawn up using environmental attributes such as soil characteristics, as was done in the coastal communities of the Monach Islands National Nature Reserve (Randall 1976). Similarly independently collected environmental attributes can be plotted on ordination diagrams produced by using the total vegetation data bank. A complete understanding of vegetational distribution in any area can only be a theoretical endpoint to research since many unrecordable environmental factors and chance play their part in every distribution.

Traditionally biogeographers have attempted to analyse the total vegetation complex leaving more limited components of it, such as individual species, to the biologist (Sinker 1964). However, the distribution and abundance of individual species may often be a key to the understanding of the way in which a whole community works. This could be the result of such factors as toxicity, nutrient uptake, or diaspore dissemination, for example. The techniques for these less comprehensive analyses of vegetation are thoroughly covered by Greig-Smith (1964) and Kershaw (1973) and will not be treated here.

Experimentation

Another area of analysis which the biogeographer has avoided is laboratory experimentation, yet this is often the only way in which hypotheses, generated as a result of field data collection, can be properly tested. In several studies carried out by the author the role of marine salts on individual species and composite field turves have been elucidated by keeping all other environmental variables constant. The effects of sand burial can also be isolated in the laboratory and access to a wind tunnel can provide useful data on evapo-transpiration. Just as with the scale models in geomorphology, however, there are considerable difficulties associated with extrapolating laboratory or greenhouse results to the real-world situation. Some geographers have denigrated biogeographical experimentation because vegetational processes are too slow. In fact a great deal of information can be gained in one growing season provided that the experiment is started at the correct time of year.

Another form of experimentation that can help to explain the spatial variations within an ecosystem is the field trial, a technique of the agronomist. Many nature reserve management bodies are presently involved in either mowing or grazing trials to assess the role of taller vegetation on orchid populations of roadside verges or the herb component of chalk grasslands. Exclosure quadrats can be erected on a semi-permanent basis with meshes of different sizes to eliminate grazers

The various analytical methods outlined in this chapter will be used according to the theoretical basis upon which the student is working and according to the aim of the project. There is no 'right' method for biogeography as a whole. Frequently techniques have been shown to have a predictive or applied role and it is within these areas that vegetation analysis best complements human geography on the one hand and physical geography on the other.

Conclusion

Thus the major reasons for the geographer studying and analysing vegetation are to complete his understanding of the major spatial phenomena on the surface of the earth, to provide a scientific ecological basis for nature conservation, to monitor the effects of environmental change such as pollution, and to appreciate the effects of man on landscape in all phases from early civilization to urbanization. Only by study at all these scales can the biogeographer help to prevent the further diminution of vegetative diversity on the surface of the earth.

References

(Asterisks suggest references of general interest or of particular value)

Acocks, J. P. H. (1953) 'Veld types of South Africa', Union Sth Africa, Dept. Agric., Div'n. Bot., *Bot. Surv. Mem.,* 28, 1–92.

Adamson, R. S. (1938) *The vegetation of South Africa* (British Empire Vegetation Committee, London).

*Anderson, D. J. (1965) 'Classification and ordination in vegetation science: controversy over a non-existent problem?', J. Ecol., 53, 521–526.

Ashby, E. and Pidgeon, I. M. (1942) 'A new quantitative method of analysis of plant communities', *Aust. J. Sci.,* 5, 19–27.

*Austin, M. P. and Orloci, L. (1966) 'Geometric models in ecology II—an evaluation of some ordination techniques', *J. Ecol.,* 54, 217–227.

Bannister, P. (1966) 'The use of subjective estimates of cover-abundance as the basis for ordination', *J. Ecol.,* 54, 665–674.

*— (1968) 'An evaluation of some procedures used in simple ordinations', *J. Ecol.,* 56, 27–34.

Barkham, J. P. (1968) The ecology of the ground flora of some Cotswold beechwoods (Ph.D. thesis, Univ. Birmingham).

Batchelor, R. and Hirt, H. (1966) *Fire in tropical forests and gresslands* (U.S. Army Lab. Natick).

Beals, E. (1960) 'Forest bird communities in the Apostle Islands of Wisconsin', *Wilson Bull.,* 72, 156–181.

*Becking, R. W. (1957) 'The Zürich-Montpellier School of phyto-sociology', *Bot. Rev.,* 23, 411–488.

Billings, W. D. and Mooney, H. A. (1959) 'An apparent frost hummock—sorted polygon cycle in the Alpine tundra of Wyoming', *Ecology,* 40, 16–20.

Boughey, A. S. (1968) *Ecology of populations* (MacMillan).

Boyce, S. G. (1954) 'The salt spray community', *Ecol. Monogr.,* 24, 26–27.

*Braun-Blanquet, J. (1965) *Plant sociology: the study of plant communities* (Hafner), Transl. rev. and ed. by C. D. Fuller and H. S. Conard.

*Bray, R. J. and Curtis, J. T. (1957) 'An ordination of the upland forest communities of southern Wisconsin', *Ecol. Monogr.,* 27, 325–349.

Brereton, A. J. (1971) 'The structure of the species populations in the initial stages of salt-marsh succession', *J. Ecol.,* 59, 321–338.

Brown, R. T. and Curtis, J. T. (1952) 'The upland conifer hardwoods of northern Wisconsin', *Ecol. Monogr.,* 22, 217–234.

Cain, S. A. (1934) 'A comparison of quadrat sizes in a quantitative phytosociological study of Nash's Woods, Posey County, Indiana', *Amer. Midl. Nat.,* 15, 529–566.

— (1947) 'Characteristics of natural areas and factors in their development', *Ecol. Monogr.,* 17, 185–200.

— (1950) 'Life-forms and phytoclimates', *Bot. Rev.,* 16, 1–32.

Carter, G. F. (1953) 'Plants across the Pacific', *Am. Antiq.,* 18, 62–71.

Céska, A. and Roemer, H. (1971) 'A computer program for identifying species-relevé groups in vegetation studies', *Vegetation,* 23, 255–277.

Chapman, V. J. (1938) 'Marsh development in Norfolk', *Trans. Norf. Norw. Nat. Soc.,* 14, 394–397.

— (1960) 'The plant ecology of Scolt Head Island', *in* Steers, J. A. (ed.) *Scolt Head Island* (Heffer).

Clements, F. E. (1904) 'Development and structure of vegetation', *Rep. Bot. Surv. Nebra.,* 7.

References

*— (1928) *Plant succession and indicators* (Wilson).

— (1936) 'The nature and structure of the climax', *J. Ecol.*, 24, 252–284.

— Weaver, J. E. and Hanson, H. C. (1929) 'Plant competition: an analysis of community functions', *Carnegie Inst. Washington, Publ.*, 398, 1–340.

Colebrook, J. M. (1964) 'Continuous plankton records: a principal components analysis of the geographical distribution of zooplankton', *Bull. Mar. Ecol.*, 6, 78–100.

Cooper, W. S. (1926) 'The fundamentals of vegetation change', *Ecology*, 7, 391–413.

— (1937) 'The problem of Glacier Bay, Alaska. A study of glacier variations', *Geogr. Rev.*, 27, 37–62.

Cottam, G. (1949) 'The phytosociology of an oak wood in south-western Wisconsin', *Ecology*, 30, 271–287.

— Curtis, J. T. and Hale, B. W. (1953) 'Some sampling characteristics of randomly dispersed individuals', *Ecology*, 34, 741–757.

*Cowles, H. C. (1899) 'The ecological relations of the vegetation on the sand dunes of Lake Michigan. I.', *Bot. Gaz.*, 27, 95–117 *et seq.*

Crawford, R. M. M., Wishart, D., and Campbell, R. M. (1970) 'A numerical analysis of high altitude scrub vegetation in relation to soil erosion in the eastern Cordillera of Peru', *J. Ecol.*, 58, 173–192.

Crocker, R. L. and Major, J. (1955) 'Soil development in relation to vegetation and surface age at Glacier Bay, Alaska', *J. Ecol.*, 43, 427–448.

*Curtis, J. T. (1959) *The vegetation of Wisconsin: an ordination of plant communities* (Wisc. U.P.).

— and McIntosh, R. P. (1950) 'The interrelations of certain analytic and synthetic phytosociological characters', *Ecology*, 32, 476–496.

Dagnelie, P. (1965) 'L'étude des communauntes végétales par l'analyse statistique des liaisons entre les éspeces et les variables ecologiques. I. Principes fondamontaux. II. Un example', *Biometrics*, 21, 345–361, 890–907.

*Dansereau, P. (1957) *Biogeography: an ecological perspective* (Ronald).

— Buell, P. F. and Dagon, R. (1966) 'A universal system for recording vegetation II. A methodological critique and an experiment', *Sarracenia*, 10, 1–64.

Daubenmire, R. F. (1966) 'Vegetation: identification of typical communities', *Science* 151, 291–298.

*— (1968) *Plant communities: a textbook of plant synecology* (Harper and Row).

Davis, W. M. (1909) *Geographical essays* (Boston U.P.), repr. 1954.

Dickinson, G., Mitchell, J., and Tivy, J. (1971) 'The application of phytosociological techniques to the geographical study of vegetation', *Scott. geogr. Mag.*, 87, 83–102.

Drude, O. (1913) *Die Okologie der Pflanzen* (Braunschweig. Vieweg Verlag).

Du Rietz, G. E. (1921) *Zur methodologischen Grundlage der modernen Pflanzensoziologie* (Holzhausen. Akadem. Abh. Wein).

Edlin, H. L. (1966) *Trees, woods and man* (Collins).

— (1973) *Woodland crafts in Britain* (David and Charles).

Edwards, A. W. F. and Cavalli-Sforza, L. L. (1965) 'A method for cluster analysis', *Biometrics*, 21, 362–375.

Edwards, K. C. (1965) 'The importance of biogeography', *Geography*, 49, 85–97.

Engler, A. and Diels, L. (1936) *Syllabus der Pflanzenfamilien* (Borntraeger).

Faegri, K. and Iversen, J. (1964) *A textbook of pollen analysis* (Munksgaard).

Fairbrother, N. (1970) 'New lives, new landscapes' (Architectural Press).

Field, W. O. (1947) 'Glacier recession in Muir Inlet, Glacier Bay, Alaska', *Geogr. Rev.*, 37, 369–399.

*Forbes, S. A. (1925) 'The lake as a microcosm', *Bull. Ill. St. nat. Hist. Surv.*, 15, 537–550.

Fosberg, F. R. (1961) 'A classification of vegetation for general purposes', *Trop. Ecol.*, 2, 1–28.

— (1965) 'The entropy concept in ecology', *in Proc. Symp. on Ecol. Res. in Humid Tropics Vegetation*, Kuching, Sarawak, July 1963 (UNESCO Tokyo).

Gittins, R. (1965) 'Multivariate approaches to a limestone grassland community', *J. Ecol.*, 53, 385–409.

References

Gleason, H. A. (1917) 'The structure and development of the plant association', *Bull. Torrey Bot. Club*, 44, 463–481.

*— (1939) 'The individualistic concept of the plant association', *Am. Mid. Nat.*, 21, 92–110.

Godwin, H. and Conway, V. W. (1939) 'The ecology of a raised bog near Tregaron, Cardiganshire', *J. Ecol.*, 27, 313–359.

*Good, R. (1974) *Geography of the flowering plants* (Longman).

Goodall, D. W. (1952) 'Some considerations in the use of point quadrats for the analysis of vegetation', *Aust. Jour. Sci. Res. Ser. B.*, 5, 1–41.

— (1953) 'Objective methods for the classification of vegetation I. The use of positive interspecific correlation. II. Fidelity and indicator value', *Aust. J. Bot.*, 1, 39–63, 434–456.

— (1954) 'Objective methods for the classification of vegetation III. An essay in the use of factor analysis', *Aust. J. Bot.* 2, 304–324.

*— (1970) 'Statistical plant ecology', *Ann. Rev. Ecol. Syst.*, 1, 99–124.

Goode, J. P. and Espenshade, E. B. Jr. (1950) *Goode's School Atlas* (Rand McNally).

Graham, E. H. (1944) *Natural Principles of Land Use* (Oxford U.P.).

*Greig-Smith, P. (1957) *Quantitative plant ecology* (Butterworth), 2nd edn. 1964.

— (1961) 'Ecological terminology', *in* Gray, P. (ed) *Encyclopedia of Biological Science* (Reinhold).

Hadfield, M. (1957) *British trees: a guide for everyman* (Dent).

Hanson, H. C. and Love, L. D. (1930) 'Size of list quadrat for use in determining the effects of different systems of grazing upon *Agropyron smithii* mixed prarie', *J. Agric. Res.*, 41, 549–560.

Harper, J. L. and Sagar, G. R. (1953) 'Some aspects of the ecology of buttercups in permanent grassland', *Proc. Br. Weed Control Conf.*, 1953, 256–65.

Hartke, W. (1951) 'Die Heckenlandschaft. Der geographische Charakter eines Land-eskulpturproblems', *Erdkunde*, 5, 132–152.

Hawksworth, D. L. and Rose, F. (1970) 'Qualitative scale for estimating sulphur dioxide air pollution in England and Wales using epiphytic lichens', *Nature*, Lond 227, 145–8.

Hayek, A. (1926) *Allgemeine Pflangengeographie* (Borntraeger).

Heyerdahl, T. (1950) *Kon-Tiki* (Rand McNally).

Hills, T. L. and Randall, R. E. (1968) *The ecology of the forest/savanna boundary* (Savanna Res. Ser., McGill Univ., Montreal).

Hooper, M. D. (1970) 'Dating hedges', *Area*, 4, 63–65.

Hope-Simpson, J. F. (1940) 'On the errors in the ordinary use of subjective frequency estimates in grassland', *J. Ecol.*, 28, 193–209.

Hughes, G. (1750) *The natural history of the island of Barbados* (London)

Humboldt, A. von (1805) *Essay sur la Géographie des Plantes* (Levrault, Schoell et Cie).

*Kellman, M. C. (1975) *Plant Geography* (Methuen).

*Kershaw, K. A. (1964) *Quantitative and dynamic ecology* (Arnold), 2nd edit. 1973.

Kitteridge, J. (1948) Forest influences (FAO, Forestry and Forestry Products, Studies).

Küchler, A. W. (1949) 'A physiognomic classification of vegetation', *Ann. Assoc. Am. Geogr.*, 39, 201–210.

*— (1967) *Vegetation mapping* (Ronald).

*Lambert, J. M. and Dale, M. B. (1964) 'The use of statistics in phytosociology', *Adv. Ecol. Res.*, 2, 59–99.

Lance, G. N. and Williams, W. T. (1967) 'A general theory of classificatory sorting strategies. I. Hierarchical systems', *Comput. J.*, 9, 373–380.

Lawrence, D. B. (1958) 'Glaciers and vegetation in south-eastern Alaska', *Amer. Scient.*, 46, 89–122.

Leopold, A. (1966) *A sand county almanack* (Oxford U.P.).

Linton, D. L. (1951) 'Vegetation', *in Oxford Atlas* (Oxford U.P.).

Loucks, O. L. (1962) 'Ordinating forest communities by means of environmental scalars and phytosociological indices', *Ecol. Monogr.*, 32, 137–166.

References

*McIntosh, R. P. (1967) 'The continuum concept of vegetation', *Bot. Rev.*, 33, 130–189.

McMillan, C. (1960) 'Ecotypes and community functions', *Am. Nat.* 94, 245–255.

MacNaughton-Smith, P. *et al.* (1964) 'Dissimilarity analysis; a new technique of hierarchical subdivision', *Nature, Lond.*, 202, 1034–1035.

Major, J. (1969) 'Historical development of the ecosystem concept', *in* Van Dyne, G. M. (ed.) *The Ecosystem Concept in Natural Resource Management*. (Academic Press).

Marbut, C. F. (1927) 'A scheme for soil classification', *Proc. 1st Internat. Congr. Soil Sci.*, 4, 1–31.

Mather, P. M. and Doornkamp, J. C. 'Multivariate analysis in geography with particular reference to drainage basin morphometry' *Trans. Inst. Br. Geogr.*, 51, 1970, 163–168.

Mead, W. R. (1966) 'The study of field boundaries', *Geogr. Zeitschr.*, 54, 101–117.

Mellanby, K. (1967) *Pesticides and pollution* (Collins).

*Moore, J. J. (1962) 'The Braun-Blanquet system: a reassessment', *J. Ecol.*, 50, 761–9.

Moore, P. D. (1971) 'Computer analysis of sand dune vegetation in Norfolk, England and its implications for conservation', *Vegetatio*, 23, 323–338.

Mueller-Dombois, D. (1965) 'Initial stages of secondary succession in the coastal Douglas-Fir and Western Hemlock zones', *Ecology of Western North America (Univ. Brit. Col.)*, 1, 38–41.

*– and Ellenberg, H. (1974) *Aims and methods of vegetation ecology* (Wiley).

Norris, J. M. (1968) *The variation of soil in some Cotswold beechwoods* (Ph.D. thesis, Univ. Birmingham).

O'Connor, F. B. (1974) 'The ecological basis for conservation', *in* Warren, A. and Goldsmith, F. B. (eds.) *Conservation in Practice* (Wiley).

*Odum, E. P. (1959) *Fundamentals of ecology* (Saunders), 3rd edit. 1971.

– (1960) 'Organic production and turnover in old field successions', *Ecology*, 41, 34–49.

O'Hare, G. (1973) 'Lichen techniques of pollution assessment', *Area.* 5, 223–229.

Olson, J. S. (1958) 'Rates of succession and soil changes on southern Lake Michigan sand dunes', *Bot. Gaz.* 199, 125–170.

O'Riordan, T. (1971) 'Environmental management', *Prog. in Geogr.*, 3, 171–231.

Orloci, L. (1966) 'Geometric models in ecology I. The theory and application of some ordination methods', *J. Ecol.* 54, 193–215.

– (1967) 'An agglomerative method for classification of plant communities', *J. Ecol.* 55, 193–205.

Perring, F. H. (1967) 'Changes in chalk grassland caused by galloping', *in* Duffey, E. (ed.) *The biotic effects of public pressures on the environment* (NERC).

*– and Walters, S. M. (1962) *Atlas of the British Flora* (Nelson), 2nd edit. 1976 (E.P.).

Peterken, G. F. (1967) *Guide to check sheet for IBP areas* (Blackwell).

Phillips, J. (1934–5) 'Succession, development, the climax and the complex organism. An analysis of concepts. Pts. I and II', *J. Ecol.*, 22, 559–571, *J. Ecol.*, 23, 210–246.

– (1935) 'Succession, development, the climax, and the complex organism. An analysis of concepts. III. The complex organism: conclusions', *J. Ecol.*, 23, 488–508.

Pielou, E. C. (1969) *An introduction to mathematical ecology* (Wiley).

Pollard, E., Hooper, M. D. and Moore, N. W., 'Hedges' (Collins), 1974.

*Poore, M. E. D. (1955) 'The use of phytosociological methods in ecological investigations. Parts I, II, III', *J. Ecol.*, 43, 226–269, 606–651.

– (1956) 'The use of phytosociological methods in ecological investigations. IV. General discussion of phytosociological problems', *J. Ecol.*, 44, 28–50.

Randall, R. E. (1968) *Vegetation zones and environment on the Barbados coast* (M. Sc. thesis, McGill Univ. Montreal).

– (1970) 'Vegetation and environment on the Barbados coast', *J. Ecol.*, 58, 155–172.

References

— (1972) *Vegetation in a maritime environment: the Monach Isles* (Ph.D. thesis, Univ. Cambridge).

— (1972) 'The origin and dissemination of the sweet potato (*Ipomoea batatas* (L.) Lam.)—review', *Trop. Ecol.*, 13, 52–64.

— (1973) 'Calcium carbonate in dune soils: evidence for geomorphic change', *Area*, 5, 308–310.

— (1976) 'Machair Zonation of the Monach Isles, N.N.R., Outer Hebrides, *Trans. bot. Soc. Edinb.*, 42, 441–462.

*Ranwell, D. S. (1972) *Ecology of salt marshes and sand dunes* (Chapman and Hall).

— Winn, J. M. and Allen, S. E. (1973) *Road salting effects on soil and plants* (NERC).

Raunkiaer, C. (1908) 'Livsformernes statistik som Grundlag for biologisk Plantgeografi', *Bot. Tidsskr.*, 29.

— (1934) *The life forms of plants and statistical plant geography* (Clarendon).

Raup, H. M. (1947) 'Some natural floristic areas in boreal America', *Ecol. Monogr.*, 17, 221–234.

*Richards, P. W. (1952) *The tropical rain forest* (Cambridge U.P.).

*Ridley, H. N. (1930) *The dispersal of plants throughout the world* (Reeve).

Rübel, E. (1930) *Die Pflanzangesellschaftan der Erde* (Huber Verlag).

Rummell, R. J. (1968) 'Understanding factor analysis', *J. Conflict Resolution*, 11, 444–480.

*Salisbury, E. J. (1952) *Downs and dunes* (Bell).

Schimper, A. F. W. (1898) *Pflanzengeographie auf ökologischer Grundlage* (Jena), 3rd edit. 1935.

— and von Faber, F. C. (1935) *Pflanzengeographie auf physiologischer Grundlage* (Jena).

Schumm, S. A. and Lichty, R. W. (1965) 'Time, space and causality in geomorphology', *Am. J. Sci.*, 263, 110–119.

Scott, G. A. M. and Randall, R. E. (1976) '*Crambe maritima* L.', *J. Ecol.*, 64, 1077–1091.

*Shimwell, D. W. (1971) *Description and classification of vegetation* (Sidgwick and Jackson).

Sinker, J. (1964) 'Vegetation and the teaching of geography in the field', *Geography*, 49, 105–110.

Sparks, B. W. (1960) *Geomorphology* (Longman), 2nd edit. 1974.

Spence, D. H. N. (1957) 'Studies on the vegetation of Shetland. I. The serpentine debris in Unst', *J. Ecol.*, 45, 917–945.

Stamp, L. D. (1969) *Nature conservation in Britain* (Collins).

Stoddart, D. R. (1965) 'Geography and the ecological approach', *Geography*, 50, 242–251.

Sukachev, V. (1945) 'Biogeocoenology and phytocoenology', *C.R. Acad. Sci. U.S.S.R.* 47, 429–431.

Tansley, A. G. (1916) 'The development of vegetation', *J. Ecol.*, 4, 298–204.

— (1929) 'Succession, the concept and its values', *Proc. Int. Cong. Plant Sci. Ithaca, 1926*, 1, 677–686.

— (1935) 'The use and abuse of vegetational concepts and terms', *Ecology*, 16, 284–307.

*— (1939) *The British Islands and their vegetation* (Cambridge U.P.), 2nd edit. 1953.

— and Chipp, T. F. (1926) *Aims and methods in the study of vegetation* (British Empire Vegetation Committee, London).

Tivy, J. (1971) *Biogeography* (Oliver and Boyd).

Van Steenis, C. G. G. J. (1969) 'Plant speciation in Malasia with special reference to the theory of non-adaptive saltatory evolution', *in* Lowe-McConnell, R. H. (ed.) *Speciation in tropical environments* (Academic Press).

Ward, J. H. (1963) 'Hierarchical grouping to optimize an objective function', *J. Am. Statist. Ass.*, 58, 236–244.

*Warming, E. (1895) *Oecology of plants: an introduction to the study of plant communities* (Oxford U.P.), 2nd edit. 1909, 3rd edit. 1925.

*Warren, A. and Goldsmith, F. B. (1974) *Conservation in practice* (Wiley).

Watson, E. V. (1968) *British mosses and liverworts* (Cambridge U.P.).

*Watt, A. S. (1947) 'Pattern and process in the plant community', *J. Ecol.*, 35, 1–22.

— (1955) 'Bracken versus heather, a study in plant sociology', *J. Ecol.*, 43, 490–506.

— and Fraser, G. K. (1933) 'Tree roots in the field layer', *J. Ecol.*, 21, 404–414.

*Watt, K. E. F. (1973) *Principles of environmental science* (McGraw-Hill).

*Watts, D. W. (1971) *Principles of biogeography* (McGraw-Hill).

*Webb, D. A. (1954) 'Is the classification of plant communities either possible or desirable?', *Bot. Tidsskr.*, 51, 362–370.

— (1953) 'A consideration of climax theory: the climax as a population and pattern', *Ecol. Monogr.*, 23, 41–78.

*— (1967) 'Gradient analysis of vegetation', *Biol. Rev.*, 49, 207–264.

*Williams, W. T. and Lambert, J. M. (1959) 'Multivariate methods in plant ecology. I. Association-analysis in plant communities', *J. Ecol.* 47, 83–101.

— and Lance, G. N. (1966) 'Multivariate methods in plant ecology. V. Similarity analysis and information analysis', *J. Ecol.*, 54, 427–445.

Wishart, D. (1969) *Fortran II programs for 8 methods of cluster analysis (Clustan I)* (Kans. Geol. Comp. Contr. 38).

Yarranton, G. A. (1967) 'Organismal and individualistic concepts and the choice of methods of vegetation analysis', *Vegetatio*, 15, 113–116.

Yemm, E. W. and Willis, A. J. (1962) 'The effect of maleichydrazide and 2,4,-Dichlorophenoxyacetic acid on roadside vegetation', *Weed Research*, 2, 24–40.

Yen, D. E. (1963) 'Sweet potato variation and its relation to human migration in the Pacific', *in* Barrau, J. (ed.) *Plants and migrations of Pacific peoples* (Bernice F. Bishop Mus. Press).

Oxford University Press, Walton Street, Oxford OX2 6DP

OXFORD   LONDON   GLASGOW   NEW YORK
TORONTO   MELBOURNE   WELLINGTON   CAPE TOWN
IBADAN   NAIROBI   DAR ES SALAAM   LUSAKA   ADDIS ABABA
KUALA LUMPUR   SINGAPORE   JAKARTA   HONG KONG   TOKYO
DELHI   BOMBAY   CALCUTTA   MADRAS   KARACHI

© Oxford University Press 1977

ISBN 0 19 874040 9

Set by Hope Services, Wantage
Printed in Great Britain
by J. W. Arrowsmith Ltd., Bristol